Antonio León Sánchez

TOWARDS A DISCRETE COSMOLOGY

Introducing a new formal basis
for a new discrete cosmology

Revised version: September 2025.

Revised version: September 2025
First edition: September 2023.
ISBN 9798861779227
All right reserved.
Copyright code SC 2303223840803
Salamanca and Santiago del Collado (Ávila), Spain.

Table of contents

Presentation

This book gathers 17 articles published between November 2022 and February 2023 in The General Science Journal. Starting from a single inductive principle (which establishes the evolution of the universe in the direction of increasing its entropy) and the inconsistency of the Hypothesis of the Actual Infinity (of which a demonstration is included in the third article), a set of theorems and corollaries is deduced, including results that prove the inevitable incompleteness of human knowledge: the need for undefinable (primitive) concepts, indemonstrable statements (axioms, and inductive principles) and unknowable causes. The elements proved are then proposed to build the formal basis for a new cosmology in which both space and time are necessarily of a real, finite and discrete nature, with indivisible minimal units, here called respectively qusits and qutits. The same formal elements (together with the concept of physical information) are also reused to resolve Kant's four antinomies. Finally, the concept of complete totality is used throughout the book and, although the union of the two words, "totality" and "completeness", is sufficiently explicit, it can also be formally defined: A complete totality is a set defined by comprehension in which every element that satisfies the law of set membership is in the set.

Paper 1. Unsolved foundational problems

Abstract.-This is the first in a series of 17 articles that aim to open a debate on the foundation of a new finitist and discrete cosmology. All the articles in the series have already been written, and will be published weekly from the last week of October 2022. The objective of this first article in the series is to expose a set of fundamental problems of logic, mathematics and physics that have not been properly resolved, and that presumably will not be, because at the base of all of them is an inconsistency directly related with the old Hypothesis of the Actual Infinity. An inconsistency that contemporary mathematics and physics insist on ignoring, despite its catastrophic consequences in the development of physical theories, particularly of cosmological theories.

Keywords: actual infinity, problem of change, infinite regress, foundation of science, laws of logic, preinertia, real and unreal space and time, irreversible time.

1.1 Introduction

I was educated first as a geologist and then as a mathematician. For that reason my clash with the actual infinity was devastating for my original scientific project (thermodynamics of organic evolution) which was interrupted forever in 1995. I was warned that the new path, the critique of infinitism, would lead me directly to ostracism. They were right[1], but I have not minded staying in academic exile for so many years. Having said that about myself,

---

[1]Therefore, thanks to The General Science Journal for publishing my work and to the Academia and Researchgate scientific networks for contributing to its public dissemination.

let's start talking about what we have to talk about in this series of articles.

There are at least seven foundational problems of contemporary science, some of them of pre-Socratic origin, that have been forgotten or not adequately addressed, despite their great relevance in disciplines such as physics and cosmology. They are the following:

1. The Hypothesis of the Actual Infinity.
2. The problem of change.
3. The infinite regress af arguments, definitions and causes.
4. The foundation of science.
5. Preinertia.
6. The real/unreal nature of space and time.
7. The irreversible nature of time.

The following sections introduce them, as well as the role they will play in the discussions developed in the subsequent articles of this series of articles. The main objective of these discussions is to propose a new foundational basis for a model of the universe whose main feature would be a finite and (above all) discrete space and time instead of the infinitist spacetime continuum of the current model. In the case of time, and apart from being finite and discrete, it would also be essentially directional and irreversible.

1.2 The Hypothesis of the Actual Infinity

The discussions about infinity are as old as abstract thought itself, which, as is well known, born in pre-Socratic Greece (though influenced by the empirical knowledge previously developed in the so called Mesopotamian river cultures [286, 28, 293, 243, 307]). In fact, the concept of (actual) infinity is already present, and in a very significant way, in Zeno's paradoxes [126, 62, 290, 289, 220, 190]. It is surely the most conflicting abstract concept created by man. And the greatest known source of paradoxes. Twenty-seven centuries of discussions were not enough to prove (or disprove) the existence of actual infinities. So, finally, that existence had to be established axiomatically at the beginning of the twentieth century by the Axiom of Infinity, one of the foundational axioms of set theory. And one of the worst misfortunes in the history of science, as will be seen in this series of articles.

Since the beginning of the 20th century, infinitist mathematics has been absolutely hegemonic and, at the same time, intolerant of dissent. But, let me recall, an axiom is just an axiom, i.e. a statement assumed without proof, and then one that can be assumed or rejected. A statement that one has the right and the duty to put to the test, particularly if it is far from obvious. Recall that the Axiom of Infinity was (more or less explicitly) rejected by authors of the intellectual stature of L. E. J. Brouwer, C. Hermite, S. Kleene, J. König, L. Kronecker, H. Poincaré, A. Robinson, L. Wittgenstein, or H. Weyl, among others. It is, on the other hand, rather ironic that set theory, the first-born creature from the Axiom of Infinity, has been the one that has finally provided the instruments to demonstrate in formal terms the inconsistency of its own infinitist foundational axiom. Among such instruments, the ω-order that will be used in the proof of inconsistency (Hilbert's machine) reproduced in paper 3 in this series of articles.

Hilbert's machine argument, inspired by the emblematic Hilbert's Hotel, was written by the author more than twenty years ago. By that time, I was already convinced of the impossibility of publishing arguments about the inconsistency of the Hypothesis of the Actual Infinity (at least by unknown authors in the field of logic and set theory, as was my case). And Hilbert's machine argument, written in a very cautious and conciliatory way, was the last one I sent to a scientific journal (not with the intention that it be published but with the intention of being informed of any possible error). As expected, the article was rejected. The editor sent me the report of its unique referee, which basically said: "the author proves one thing and later the opposite." I wrote to the editor (mainly to check his reaction) saying that, indeed, I demonstrated a thing and then the opposite because I was demonstrating a contradiction. His response was also what was expected: no response. The reader of Part 3 will be able to decide if that argument deserves, or not, a little more consideration and respect.

The inconsistency of the actual infinity will have enormous consequences in physics (and naturally in mathematics itself) because physics has been written with infinitist mathematics for more than a century. One of these consequences is that all physical concepts and quantities (including space and time) must be discrete, with indivisible minima, as, on the other hand, it is already assumed for quantum magnitudes, although quantum physics, mainly the physics of discrete entities, is also expressed with the indiscrete mathematics of the infinitist continuum $\mathbb{R}$ of the real numbers. The same can be said of the mathematical language based on Hilbert space on which quantum field theory is built.

It does not seem an exaggeration to state that the pre-Socratic discovery of the irrational numbers and the proposal and acceptance of the Axiom of Infinity have had dire consequences in the development of mathematics and of the physical theories most committed to mathematics. Nowadays, practically nobody within the academic orthodoxy has stopped to think about the consequences of taking the continuum of real numbers as a model. Infinity theories have been growing for more than a century and a half. And nobody knows how and when they will stop growing. Let us hope that Henry Poincaré was right and the day will soon come when we will be able to say that the actual infinity was a disease of which we have already been cured [231, p. 121], [65, p. 1].

Fortunately, all experimental sciences are finitist in their results (there are no real measurements with an infinite number of decimals), and experimental results prevail over theoretical constructs. We have physical models and theories that explain the physical world reasonably well (even very well), but they are not complete theories precisely because they are founded on the infinitist hypothesis of the spacetime continuum, in which both space and time have a continuous, non-discrete nature. In this series of articles we will have the opportunity to demonstrate that neither space nor time can be divided infinitely. Therefore, they must be discrete, with indivisible minima. And this conclusion changes everything.

And not only in mathematics and physics, the inconsistency of the Hypothesis of the Actual Infinity also has consequences in logic and philosophy. In particular, the article 16 in this series of articles demonstrates how the inconsistency of that hypothesis is sufficient to resolve three of Kant's four antinomies. In the same article the solution of the fourth antinomy is also demonstrated, although now using the concept of physical information (from my original scientific research on organic evolution [188, 189]).

1.3 The problem of change

Also of pre-Socratic origin, the problem of change is surely the most difficult problem ever posed by man. So difficult that it remains unsolved more than twenty-seven centuries after it was posed. So difficult that some classic and modern authors, such as Parmenides, Zeno, McTaggart or Hegel, have defended that it could be an inconsistent process, despite its overwhelming evidence: change is the most pervasive characteristic of our continuously evolving universe [234, 291, 25, 26, 223, 247, 224, 134, 142, 331].

So, without solving the problem of change it will be impossible to explain the physical world. It is therefore surprising how little interest modern physics, the science of change, takes in the problem of change. I have the impression that many physicists ignore that the problem of change is still an unsolved problem, ignore that until now nobody has been able to explain how a simple change of position occurs.

Naturally, physics has explained a great variety of changes of all kinds (mechanical, electrical, magnetic, thermal, etc.). It has explained which constants and which variable magnitudes are involved in each of these changes, and what mathematical relations exist between these variables and these constants, but none of these changes has been explained, as such a process of change. In paper 6 of this series of articles it will be proved that change is indeed inconsistent in the spacetime continuum. But it will also be proved there that the problem of change can be solved within certain theoretical frameworks (similar to cellular automata) in which space and time are of a discrete nature.

In addition, in these discrete and finite frameworks, most of the oddities of relativity and quantum physics could be explained. Rarities that surely appear because of the insistence of physics to explain the discrete physical world by means of inappropriate continuous mathematics, the same mathematics that makes it impossible to solve the problem of change. As an anticipated conclusion, we could say that if the Hypothesis of the Actual Infinity were inconsistent (and it will be proved that it is), then space and time can only be discrete. And in a discrete space and time the problem of change can be finally solved.

1.4 Infinite regress of arguments, definitions and causes

Since statements do not prove themselves, to prove a statement, say S_1, we need at least another different statement S_2 justifying S_1 in formal terms; for the same reason, to prove S_2, we need at least another different statement S_3 justifying S_2 in formal terms; for the same reason again, to prove S_3, we need at least another different statement S_4 justifying S_3 in formal terms; and so on and on. This is the Aristotelian infinite regress of arguments, the reason for which we need axioms in formal sciences and inductive principles (or fundamental laws) in experimental sciences.

We have the same problems with definitions: since concepts do not define themselves, to define a concept, say C_1, we need at least another different concept C_2 defining C_1 in semantic terms; for the same reason, to define C_2, we need at least another different concept C_3 defining C_2 in semantic terms; and so on. This is the reason for which primitive (undefined) concepts are inevitable and necessary in all languages, either formal or ordinary.

Most basic scientific concepts are primitive: set, number, point, force, mass, energy, time, instant etc. Axioms, principles, fundamental laws and primitive concepts drastically limit human knowledge, although we tend to pay little attention to these inevitable restrictions, at least in comparison with the attention we pay to certain famous theorems of logic based on another idiotic concept: self-referent statements ("this sentence is false", and the like) [192, 194].

The need to use primitive concepts is surely behind the lack of rigor in the use of ordinary language and also of formal language that will be analyzed in the article 2. A very serious case, as will be seen in this series of articles, is the misuse of successiveness in the case of points in space and instants in time, in which there is no immediate successiveness (adjacency): the natural number 5, for example, is the immediate successor of the natural number 4, but no point (instant) can be the immediate successor of another point (instant) in the spacetime continuum.

In spite of this, the (primary and secondary) physics literature are replete with dramatically erroneous expressions such as:

   ... it propagates through the adjacent points...

   ... is distributed point to point...

   ... through each of the contiguous points...

   ... in the next instant...

   ... in the previous instant...

     etc.

Always forgetting that in the spacetime continuum there is no adjacency: between every two points (instants), whatever they are, there are always other $2^{\aleph_o}$ different points (instants). So the description of all supposedly continuous physical phenomena can only be discontinuous, in jumps. And the reason is the same reason why it is impossible to solve the problem of change or Zeno's paradoxes: the infinitist topology of the spacetime continuum.

But the infinite regress that interests us most here is that of origins (causes). Indeed, things do not originate by themselves either. The reader will be able to guess what will be found in paper 15 devoted to this subject, a subject unattended by modern science surely because of certain religious prejudices.

1.5 The foundation of science

Although the infinite regress of arguments, definitions and causes is assumed implicitly, it should be made explicit (as will be done in this series of articles, particularly in the article 8). The consequences will be very positive, making it possible to sort out the certain disorder that still exists today in the foundations of the different sciences, whether formal or experimental. Two levels of foundation should be distinguished, a general logical level that only affects the consistency of the arguments, and a particular level of each science.

At the general level, the laws that every argument must comply with have been established since the time of Aristotle [202]: at least the Law of Identity and the Law of Non-Contradiction (the Law of the Excluded Middle could also be included). As is known, these laws are assumed by all sciences and allow to establish the basic modes of inference (Modus Ponens and Modus Tollens). At the particular level, it is necessary to establish the basic laws of each science. In the case of experimental sciences, we will call them principles, they should be inductive in nature, based on observation and experimentation.

It is interesting to note at this point that the nineteenth-century naturalistic Principle of Actualism Uniformism says the same thing as the principle of relativity, although without reference to reference frames.

Natural laws are the same in all places and times.

Well, the foundation that will be proposed in this series of articles includes an inductive principle valid for all sciences and which is even more general and basic than the above mentioned Principle of Actualism Uniformism. We will call it the Principle of Directional Evolution:

The observable universe evolves independently of its human observers and always in the same direction of increasing its global entropy.

From this principle, which is explained and empirically justified in the article 5, several theorems will be proved (including the Principle of Relativity) that can be used in a new foundation of experimental sciences. Among them the following:

1. **Theorem of the Consistent Universe**: The universe evolves under the control of a unique set of invariant and consistent physical laws.

2. **Theorem of the Formal Dependence**: No concept defines itself; no statement proves itself; no physical object is the cause of itself; and no cause is the cause of itself.

3. **Theorem of Reference Frames**: The laws of physics are the same in all reference frames.

4. **Theorem of the Extensive Intervals**: The intervals of space and time within which the physical laws apply must always be greater than zero.

5. **Theorem of the Indivisible Units**: There is an indivisible minimum of space (time) intervals of which all space (time) intervals are an integer multiple.

6. **Theorem of the Discrete Threshold**: Physical laws do not apply in length intervals less than the quantum of length nor in time intervals less than the quantum of time.

7. **Theorem of Adjacency**: No space exist between any two successive quanta of space, and no time elapses between two successive quanta of time.

8. **Theorem of Preinertia**: Every physical object inherits in one of its vector components the relative velocity vector of the reference system where it is set in motion, provided that the resulting speed does not exceed the possible maximum limit.

9. **Theorem of the Discrete Motion**: The continuum densely ordered spacetime cannot be used to model uniform motion.

10. **Theorem of Physical Space and Time**: The indivisible units of space and time are physical, and then real and absolute.

11. **Theorem of the Arrow of Time**: In a consistent universe the joint evolution of any system and its environment is always in the same direction of increasing its entropy.

1.6 Preinertia

In this case, preinertia it is not an inconsistency, nor an unsolved and forgotten problem, nor a formal impossibility. It is a concept assumed by modern physics, although only implicitly, surely because no relevant physicist has ever thought of the real existence of this universal property of all physical objects. And here is the only inconvenience, that concept should be explicitly declared in order to use it in the construction of models and theories on the physical world. Preinertia is the ability of all physical objects to inherit the *relative* velocity vector of the inertial reference frame in which they are set in motion, including photons created and set in motion in any reference frame.

Special relativity, for example, would be impossible without preinertia. Indeed, assume a photon a^* is emitted by its source S in the direction parallel to the axis Y_o of the proper inertial reference frame RF_o of the source S. If RF_v is another inertial reference frame that coincided with RF_o at a certain instant and from whose perspective RF_o moves with a velocity v in the direction of the increasing axis X_v of RF_v, the photon a^* will be observed moving along a trajectory inclined by an angle α respect to the axis Y_v of RF_v. It can be easily proved that the vector components of its velocity are:

$$c_y = c \sin \alpha \tag{1}$$

$$c_x = c \cos \alpha = v \tag{2}$$

So that for RF_v-observers, the photon a^* inherits the relative vector velocity $\vec{v}$ of the inertial reference frame of its emitting source. In a more general sense, it can also be proved that all physical objects, including massless objects as photons, are preinertial (paper 7 in this series, and [197]).

Being a universal and significant attribute of all physical objects supported by the highest empirical evidence (confirmed, for example, every time an object falls on the surface of the Earth: it falls vertically from the point at which the fall begins), preinertia could be included in the statement of the Principle of Inertia as we will do in paper 7:

> Every physical object is preinertial and remains at rest or moves at a constant uniform velocity, unless an external force acts upon it.

It remains to analyze the role of this universal preinertia in gravitational attraction. And it could be the case that such a role would greatly simplify the explanation of the observed geodesic curvatures, even without the need for the cumbersome deformation of the spacetime continuum, an infinitist concept that could be inconsistent (paper 3 in this series proves that is the case), and is usually considered in modern physics as unreal. By the way, how could an unreal object vibrate -gravitational waves- and deform? (more on this subject in the following articles of this series).

1.7 The real/unreal nature of space and time

A major (though ignored) problem of contemporary physics is the division between the supporters of the unreal nature of space and time and the supporters of the real nature of both physical entities. This is indeed a serious problem, but a problem that is not even posed, as if it did not matter in the slightest. In this series of articles, particularly in the articles 9 and 10, we will have the occasion to see that it does matter, which will become evident as soon as the right questions are asked.

The supporters of the illusory character of space and time are in the majority with respect to the supporters of their physical reality. And both use as their basic model the set of real numbers, which is densely ordered, an ordering that does not allow immediate successiveness (adjacency): between any two real numbers there is always a non-numerable infinity of other, different real numbers. Under these conditions it is impossible to describe most physical events, for the same reason that it is impossible to solve Zeno's Dichotomy or the problem of change. In this sense, physical theories, as such theories, would be incomplete. But the vast majority of physicists do not even consider these questions.

The proponents of the non-real nature of space and time would have to explain, in addition, how something that does not exist, that is only an illusion, can extend, deform vibrate and transmit its own vibrations (gravitational waves). They would have to explain how it is possible that physical objects can move following non-existent geodesics, non-existent because the mathematical space in which they are defined does not physically exist. Do they move through equations or through some kind of reality?

1.8 The irreversible nature of time

Although each of us has a personal experience of time, and the nature of time has always been an important object of scientific discussion, until now it has not been possible to define neither the concept of time nor that of instant. They are, therefore, primitive concepts. St. Augustine's well-known phrase about what time is sums up our experience with it very well [67, XI, 14, 7, p. 560]:

> If no one asks me, I know; if someone asks me and I want to explain it, I don't know.

In his Critique of Pure Reason, Kant deeply analyzed the concept of time and his analysis has been very influential in modern science. For Kant, time is not a property of matter, it does not exist as an independent physical object. It is an instrument of human perception; a relational instrument that our mind uses to explain the world in which it operates [171, B 54, p.165]:

> Time, therefore, is not to be regarded as an object, but as the mode of representation of myself as an object.

And according to Boltzmann's statistical interpretation, time flows in one direction (arrow of time) because it is vastly more likely to do so: the sense of time is a statistical property. Therefore, the fact that time flows backwards is very unlikely, but not impossible. Time would therefore be reversible. Physical theories, almost without exception, are compatible with a reversible time, which in addition is usually interpreted as relational, devoid of absolute existence. Nor does absolute time exist in the theory of relativity. And Gödel, who in the last years of his life also dealt with time, defended ideas very similar to those of Kant [132].

For many contemporary authors time does not exist, it is a mere consequence of covariant quantum fields [280]. But the stratigraphic series of sedimentary rocks are there, arranged from top to bottom so that each layer was deposited before the one above it (Law of Superposition); and its enormous macro and (above all) micropaleontological content is indicating an indisputable arrow of time that has been operative on this planet for at least the last 3600 million years. These stratigraphic series and their organic content are the indelible mark of the irreversible passage of time on Earth, in a reality that already existed as such a reality billions of years before there were human observers to observe it. A reality, then, objective and independent of human observers that has left billions of proofs for those who want to analyze them. By the way, the gasoline that moves the cars of time-denying physicists is also a proof of the irreversible passage of time.

In the article 10 of this series of articles it will be proved that, if the universe is consistent (an it will also be proved it is), time cannot be divided infinitely. Therefore, there must be indivisible minimums of time. And, as we shall see, this changes everything.

Paper 2. Physics and language

Abstract.-This article points out the frequent lack of rigor in the use of ordinary language in contemporary physics and its lack of concern for the problem. The ambiguous, even contradictory, use that physicists make of certain terms, such as order, organization, information, entropy, infinity, point, change, nothingness and others, stands out. This lack of linguistic and formal rigor leads some physicists (even Nobel Prize winners) to construct fallacious arguments. The article also highlights the disinterest of physics in the formal problems at the base of its mathematical language. Particularly noteworthy is its disregard for a hypothesis, which, if inconsistent, would have devastating consequences on physical theories: the Hypothesis of the Actual Infinity subsumed in the Axiom of Infinity, a hypothesis that underlies the spacetime continuum on which most of such physical theories are built.

Keywords: order, organization, information, entropy, infinity, actual infinity, nothingness, vacuum, change, preinertia, infinite regress, arrow of time, fabric of spacetime, theoretical physics, experimental physics.

2.1 Introduction

The aim of this second article in the series is to show the lack of rigor of contemporary physics in the use of certain key words, and its disregard for the formal foundations of its mathematical language. Despite the disastrous consequences that the inconsistency of the Hypothesis of the Actual Infinity would have on theoretical models and physical theories. It is surprising that, being the Axiom of Infinity an axiom by no means self-evident,

15

and having the Hypothesis of Actual Infinity subsumed in that axiom a long and conflicting history of more than twenty-seven centuries, practically nobody has bothered to question its formal consistency, at least in the last 100 years.

In fact, physicists spend all their time and money researching and discussing models and theories all based on the spacetime continuum, but they do not spend a single second discussing the formal consistency of that spacetime continuum. But if actual infinity is inconsistent, then so is the spacetime continuum and all the theories that use it. In any case, it is fortunate that experimental physics (always discrete and finitist) has the last word. Although the proofs are almost ostracized at the moment, the proofs of the inconsistency of the Hypothesis of Actual Infinity exist [198]. In the article 3 of this series, the reader will have the opportunity to evaluate one of them.

After a brief introduction to ordinary language and formal language, this article analyzes, also briefly, the use that physics makes of certain key terms of ordinary language such as order, organization, information, emptiness, nothingness, infinity, etc. A use that can be confusing, abusive, arbitrary and even contradictory. Such abuses of language would be unimportant if it were not for the fact that they facilitate the forgetting of certain basic problems of physics and the wrong approach to other matters. We will see some cases in this article, particularly the concepts of point and instant. Next, the disinterest of physicists in the foundations of the formal language of physics (infinitist mathematics) is made evident. Presumably they are aware that the infinity of their mathematics is the actual infinity, not the potential infinity. Although it does not always seem to be the case.

Occasionally, some aspects of the discrete and finitist alternative are also anticipated, which could serve for the creation of a new foundational basis for cosmology, which, being finitist and discrete, would be much simpler and more physical than its infinitist current alternative. Obviously, a detailed foundational basis is not proposed here, but some of the formal elements that would have to be part of it in order to be consistent and compatible with experimental observations, which is not an obstacle because all experimental observations are finite and discrete.

Cellular Automata Like Models (CALMs) will be used as a possible tool for discussion, both theoretical and practical. It also happens that at least one of the conversion factors between the discrete geometry of CALMs and the geometry of the spacetime continuum is precisely the relativistic Lorentz factor. This coincidence could make the special theory of relativity (an infinitist theory of spacetime continuum) unnecessary, or a theory of appearances experimentally confirmed [197].

2.2 Ordinary language

Living beings are the only known natural objects that exhibit arbitrary, even extravagant, properties (such as singing, or dancing, or having red feathers, or yellow feathers,....), properties that cannot be deduced directly from the physical laws but from an evolutionary competition driven by the powerful force of reproduction [189]. Indeed, living beings are subject to a law that dominates all others (including physical laws):

Reproduce as you might.

Excellence in compliance with the physical laws is of no use if you do not reproduce in the end. Hence the extravagances, the arbitrary properties such as walking or crawling or jumping, are real. In the biosphere, everything that can vary, varies.

An exclusive capacity of living beings is their ability to communicate with signals that are almost always sound, visual, or chemical. With human language, this capacity reaches a level far superior to that of other animals. Language allows us to represent and communicate our thoughts and feelings. This language, which we call ordinary, evolves much faster than man himself. Unfortunately, it does not always go in the best direction. Most humans make errors in the use of the established rules for the use of the language they speak, errors that tend to generalize and become new rules of that language. It could be said that the mistakes of those who make the most mistakes mark the direction of the evolution of languages [120].

Consequently, languages become full of irregularities, they become almost chaotic. In some cases, even the original correspondence between graphic and phonetic signs has been lost: graphic signs are pronounced differently depending on the words they are part of. So it is necessary to learn how each word is written and pronounced. They are non-phonetic languages. Learning them well is almost impossible, except as a mother tongue. The initial evolution towards a non-phonetic language can be observed in Andalusian, from my homeland, Andalucía (Spain). A very advanced step in that non-phonetic direction is English. In the opposite direction we have the Spanish of, for example, Castilla y León (where I live). Therefore, the more or less chaotic use of ordinary language is normal, and, as we shall see, scientists are no strangers to this habit.

But language can also be put to formal use, even formalized in a more or less abstract way. As noted in paper 1, abstract thought begins in pre-Socratic Greece, although with antecedents in the fluvial cultures of the Near East [286, 28, 293, 243, 307]. And with it, the formalization of language was born. Euclid's Elements [94, 140, 138] is the first notable example of a formal text in which its main elements are already recognized: definitions, axioms, and deduced propositions. To these three elements must be added the laws of logic and the basic modes of inference (neither of which are explicitly stated in the Elements). As is well known, the fifth of Euclid's geometric axioms is out of tune with the rest of the axioms, and the definitions are very unproductive (they are barely used in the axioms and proofs).

Changing the definitions and the axioms we will obtain different results with the same laws of logic and the same inference engines. Hence the importance of choosing an appropriate foundational base of definitions and axioms. For example, by changing somewhat the foundational basis of Euclidean geometry, new non-Euclidean geometries are obtained [270, 340, 123, 72, 242, 275]; or new Euclidean geometries [257, 258, 144, 193]. It could be said that formalized languages such as mathematics or geometry allow the (apparently indefinite) extraction of the information packed in its foundational base. And as long as no contradictions are extracted, we can trust the veracity of that foundational basis.

The Axiom of Infinity is one of the axioms that underlies set theory, and therefore a good part of contemporary mathematics, which is also the mathematics with which modern (theoretical) physics is built. In paper 3 of this series it is proved that the infinity of the Axiom of Infinity is the actual infinity (Theorem 3.1) and that the actual infinity is inconsistent (Corollary 3.3). The rest of the articles in this series discuss very important consequences of this inconsistency in theoretical physics.

2.3 Formal language

The maximum degree of formal abstraction is reached in mathematical equations, in which symbols replace words. There is nothing new here, symbols denote previously defined concepts (or primitive concepts), all of them legitimated by the corresponding axioms or by previously demonstrated propositions. The novelty is that some of these objects are variables and are related to other objects that are also variables. But mathematical equations are also obtained by applying the laws of logic and the methods of logical inference to formal objects. The equations represent formal statements in a very compact form, i.e. expressed with relatively few symbols. But not all equations are the same, some are very simple, some are very complex and solving them is only possible through approximations made with the help of large computing resources (this is the case of many physics equations).

Mathematical equations contain a lot of information (which, by the way, can be true, false, or inconsistent). You could say that they are a very compact way of speaking. But they must be interpretable in terms of ordinary language. Otherwise they would be completely useless. The claim of some authors that certain things can only be expressed in mathematical language is not, in my opinion, correct, unless they represent unintelligible nonsense. Bertrand Russell said that in mathematics we never know what we are talking about, nor whether what we are talking about is true [284, p. 959] [287], but that was an ironic way of speaking. What was not ironic was the assertion of physicist R. Feynman that mathematics is not a science [106, v. 1].

On the other hand, it is a widespread custom among mathematicians and physicists (much less so among naturalists) to refer to the beauty of mathematical equations. As if beauty gave them a plus of veracity. I confess my inability to appreciate the aesthetic differences between, say, $2\pi^2 < 3\pi^3$ and $2\pi^2 > 3\pi^3$. It will be interesting to know whether the pretended beauty of certain mathematical equations remains such beauty once they are shown to be inconsistent. What will happen sooner or later with the infinitist mathematics (which we have let grow too long) that models the most representative physical theories of contemporary physics. Can inconsistency be beautiful? In addition, aesthetics is a personal matter (for me a poem by San Juán de la Cruz, Gregorian chant, a Beethoven sonata, and many other things that do not include mathematical equations, are beautiful).

One of the most relevant aspects of formal languages is their intended Platonic reality: mathematics would be real because it explains the physical world. This is what is claimed by the so-called indispensability argument [261], which could be summarized in the following (somewhat crudely) way:

> Mathematics is indispensable for experimental sciences (physics, for example) that satisfactorily explain objects and natural phenomena and make accurate predictions. It has to be accepted then that mathematics is true and therefore that its entities *must exist.*

It should not be forgotten that mathematics has also been (and continue to be) indispensable for constructing erroneous explanations of the world. And that does not imply that we have to declare them false, except for the alleged initial assumptions responsible for the errors. The indispensability argument has received numerous criticisms [305, 2, 3, 43], but in spite of them it remains one of the pillars of contemporary mathematical Platonism.

According to the Platonic faith, the formal use of language could even (fallaciously) prove the existence of God, for example the argument of Saint Anselm of Canterbury [66] (italic is mine):

1) Conceptually, God is the greatest being imaginable.

2) God exists as an idea in the mind.

3) A being that exists in the mind and in reality is greater than a being that exists only in the mind.

4) If God exists only in the mind, then we can imagine something greater than God. [*The problem is that anything we can imagine also exists only in the mind.*]

5) It is contradictory to imagine something greater than God, because God is the greatest being imaginable.

6) Therefore, God exists.

But in the mind there are also absurd and contradictory ideas of whose existence we do not have, nor do we hope to have, news. In fact existing in the human mind could also be considered an imperfection.

2.4 Physics and ordinary language

Physics has never had to deal with nomenclature problems, as chemistry did in order to become a valid scientific discipline [196, 24, 235]. And the same thing happened with the natural sciences (biology and geology). Physics does not have problems of nomenclature, but it does have problems with the careless use of certain key terms. An inheritance of that human propensity to make mistakes and inaccuracies in the practice of ordinary language that we saw in the previous section. But in the case of a science like physics such errors and inaccuracies should not be permissible.

Indeed, in physics there are problems related to the use of certain key terms, at least those mentioned below. They are important problems, but physics does not even consider them to be problems. It goes on its way without paying the slightest attention to them. Among the most important of these terms we must highlight the following:

1. Order and organization.
2. Entropy and information.
3. Something and nothing.
4. Point and space.
5. Instant and time.
6. Potential infinity and actual infinity.

The first two pairs of terms denote very important properties of physical objects. The following three pairs denote very significant physical entities in the development of physical theories. The last, and most important in this series of articles, represents a historical duality, poorly resolved by mathematics, and ignored in physics, which makes uncritical use of the solution axiomatically assumed by contemporary mathematics (Axiom of Infinity).

On the physical notions of order and organization I have dealt elsewhere in some detail [196]. It is, in short, the use of two terms to denote two completely different properties of many physical objects, but with the peculiarity that the two terms are used synonymously to denote both properties. Or in other words, instead of denoting each of the two distinct properties by a different name, the two distinct properties are denoted by the same two words. There is no greater nonsense. The reason for this anomaly is that physics ignores the existence of organized objects in the Kantian sense.

Organized objects were defined by Kant in his Critique of Judgment [169, $62-68] as teleological objects; objects with a natural (organisms) or artificial (machines) purpose or finality ([196, p. 6]):

> In organized objects, and only in them, the parts are
> arranged in relation to each other in a totality such

that the desired end is attainable. In the case of machines, the purpose is imposed by the intention of the man who designed them. In the case of organisms, the problem was precisely to explain their origin, the authorship of their own purpose. Kant defines organisms as natural purposes, self-organizing objects in which each of its parts exists exclusively in relation to the others.

Obviously, the natural purpose of the living being is self - reproduction. Ignoring the existence of organized and self-organized objects is the main reason why living beings remain strange objects for physics, objects difficult to explain in physical terms [292, 39, 90, 259, 119, 229, 249, 248, 250, 251, 91, 219]. But they are just as real, that is, just as physical, as any other physical object.

The situation with the term "order" is less dramatic: we would have two types of physical order: the order based on periodic regularities (crystal lattices, for example) and the thermodynamic order: a measure of the number of microstates compatible with a certain macrostate. The solution to the confusing use that physics makes of the terms order and organization is, therefore, very simple: recognize the existence of organized objects and call them organized objects; and then call ordered to ordered objects.

The problems posed by the pair of terms entropy / information are similar to those just discussed for the order / organization pair. In this case, physics only considers the probabilistic aspects of information (Shannon's information, [298]), and forgets the much more physically significant aspects that physical information represents: the ability to produce arbitrary changes; changes permitted by the laws of physics but not driven by the laws of physics but from codes arbitrarily established; changes that make possible the appearance of arbitrary qualities, not deducible from physical laws, in informed systems such as living beings (see [189]).

The novelty of these changes, allowed by physical laws but independent of them, and their importance in the evolution of the universe are so great that there should be a branch of physics dedicated exclusively to their study. Instead, physics ignores them, and uses the terms information and entropy as if they were synonyms. But physical information is a real capability of many real physical objects. Its existence explains, for example, that I am writing this, and that someone will end up reading it, recognizing that the latter is highly unlikely since I wrote and published the same thing in 1996 [189].

The next four pairs of terms listed above (something and nothing, point and space, instant and time, potential infinity and actual infinity) will be discussed in subsequent articles in this series. They form an essential part, especially the last couple, of the discussion that will justify the need to consider new discrete and finitist alternatives that consistently explain the observable physical universe.

As a simple illustration of what will be discussed below, consider the use of the terms *something* and *nothing* by an eminent contemporary physicist [177, p. 7 Kindle Ed.]:

> **PO**: *Nothing* is in everything as material and physical as *something* is, especially if it is defined as the *absence of something.*

what is said (written) with the intention of being able to affirm later that universes can be born spontaneously from nothing, a very frequent affirmation in the literature of contemporary physics. But *nothing* can have no capacities or properties, otherwise it becomes *something* with those capacities or properties. For example, *something* with the ability to spontaneously create universes. Actually, and with the same semantic right, **PO** can be paraphrased, for example, as:

> *Dead* is in everything as material and living as *living* is, especially if it is defined as the *absence of living.*

For their part, and being primitive concepts, the concepts of *object*, *space* and *time* will be used as such primitive objects in the demonstrations. Informally, and by way of illustration, the following definitions could be considered

Definition 2.1 (of Physical Object) *Physical object: any object for whose existence there is empirical evidence.*

Definition 2.2 (of Space) *Space: an extensive object that contains all other physical objects in the universe, acting also as a mediator in all their interactions at a distance.*

According to the CALM model that will be proposed in this series of articles, the previous (informal) definition of space would have to be completed by indicating that it is the physical space that creates its own objects:

Definition 2.3 (of Physical Space) *Physical space: an extensive object that contains and generates all other physical objects in the universe, acting also as a mediator in all their interactions at a distance.*

Definition 2.4 (of Time) *Time: a magnitude that measures the persistence of physical objects.*

The real, or unreal, nature of space and time will be discussed in other subsequent articles of the series.

2.5 Physics and formal language

The formalism of physics is the formalism of contemporary mathematics. A discipline in which the Hypothesis of the Actual Infinity plays a very relevant role, as will be seen throughout the articles of this series of articles. The most outstanding aspect of the relationship between the two sciences is that, on the one hand, physics makes uncritical use of mathematics; and on the other, that mathematics legitimizes the Platonic existence of its formal objects based on their productivity in describing the physical world (see the indispensability argument above).

By assuming the Axiom of Infinity, one is assuming that the incomplete (e.g. the ordered list of natural numbers, in which there is no last element that completes the list) exists as a complete totality (the set of the TOTALITY of natural numbers), were a complete totality is a set defined by comprehension in which every element that should be in the set, is in the set. Physicists seem to agree with that assumption, because that assumption underlies the mathematical language of contemporary physics. They agree, therefore, that a linear segment of Planck's length ($\approx 1.6 \times 10^{-32}$ millimeters) has as many points as the whole observable three-dimensional universe. And, therefore, that in a segment of Planck's length the same number of virtual particles is produced as in the whole observable three-dimensional universe.

Indeed, the set $\mathbb{R}^4$ of all 4-tuples models the spacetime of physics formally justifies the above assertion. But what if the actual infinity subsumed in $\mathbb{R}$ were shown to be inconsistent? Could a theory using inconsistent objects be consistent? What consequences would that inconsistency have on special relativity, a theory of the spacetime continuum? What consequences would it have on the mathematics of the Standard Model? Is it not risky to insist on marching along the same infinitist path without doing the slightest research on its possible inconsistency? Is there any finite and discrete alternative to the spacetime continuum compatible with the experimental data? These, among others, are the questions that have motivated this series of articles.

Paper 3. Finite versus infinite

Abstract.-Once made the necessary distinction between the actual infinity and the potential infinity, this paper 3 of the series proves that the infinity of the Axiom of Infinity can only be the actual infinity, and that ω-ordered collections are inconsistent, which in turn implies the inconsistency of the Axiom of Infinity itself, and the inconsistency of any infinite set when considered as a complete totality. It is also shown here (and the results will be much used in subsequent discussions) that every set is either discrete or can be ordered discretely (where being discrete means having a first and a last element and that every element (except the first) has an immediate predecessor, and an immediate successor (except the last). Obviously, the infinitist mathematics of modern physics (rarely put to the test) will be seriously affected by the inconsistency of the actual infinity (fortunately experimental physics can only be finitist and discrete). The consequences of this conclusion will be deduced in this and the following articles of the series. Here, one of such physical consequences will be demonstrated: in a consistent reality only a finite number of universes (if more than one) could exist, each with a finite number of physical objects.

Keywords: actual infinity, potential infinity, Dedekind definition, axiom of infinity, Hilbert's machine, inconsistency of ω-order, inconsistency of the actual infinity, inconsistency of actual infinite sets, theorem of the finite universe.

3.1 The actual and the potential infinity

(This section includes published texts by the author [198, p. 32-35])

In common parlance, the word infinite is used to refer to the quality of being immense, gigantic, unlimited, etc. C. F. Gauss (*Princeps Mathematicorum* [343, p. 1188]) said that infinity is a way of speaking (C.F. Gauss, Letter to astronomer H.C. Schumacher, 12 July 1831 [116, Vol. II, p. 268]):

> I protest against the use of infinite magnitude as something completed, which is never permissible in mathematics. Infinity is merely a *façon de parler* [a way of speaking], the true meaning being a limit which certain ratios approach indefinitely close, while others are permitted to increase without restriction.

The consideration of an infinite magnitude (or an infinite sequence, for instance of numbers) as something completed is what we call *actual infinity* since Aristotle, who introduced the distinction between the potential infinity and the actual infinity [14, 15, Books III, VIII]. It is remarkable the fact that in the above quotation, Gauss implicitly includes the distinction between both infinities (see below in this section). I have the impression that most physicists think, like Gauss, in terms of the potential infinity, without worrying about the fact that they are building physics with the mathematics of the actual infinity.

As will be seen below, it is possible to give a precise definition of the concept infinity, albeit based on a primitive concept: the concept of set. However, the concept of set could be defined in operational, non-Platonic terms [198, p. 360]:

Definition 3.1 (of Set) *A set is the theoretical object that results from a mental grouping of different arbitrary objects previously defined.*

This definition has the advantage of avoiding the entanglements caused by self-reference, simply by requiring that the elements to be grouped be previously defined, which seems quite reasonable if we intend to know what we are grouping. It is convenient, on the other hand, to consider that some sets exist as complete totalities, i.e. as sets satisfying the following:

Definition 3.2 (of Complete Totality) *A complete totality is a set defined by comprehension in which every element that satisfies the corresponding membership definition of the set is in the set.*

In consequence, to a complete totality of a certain type of elements, it is not possible to add new elements of that type because it already contains *all of them*.

But returning to the concept of infinity, and apart from Gauss's opinion, the word "infinite" has also a precise meaning based on the primitive concept of set:

Definition 3.3 (of Infinite Set) *A set is said infinite if it can be put into a one to one correspondence with one of its proper subsets.*

which is the well known Dedekind's definition of infinite set [68, p. 115], an important element of the foundations of modern infinitist mathematics, which began its development at the end of the 19th century. As is well known, the controversial history of the (philosophical and) mathematical infinity has its roots in the pre-Socratic times, although here we are not interested in the details of that history (there is an abundant and excellent literature on the history of infinity, for instance: [344, 212, 295, 30, 278, 60, 205, 230, 232, 180, 181, 1, 233, 231, 57, 333, 19, 276]).

Now I will try to explain the distinction between the two infinities, the actual and the potential. The set of the natural numbers and *supertask theory* are two suitable instruments to evidence such a distinction. The set of the natural numbers needs no presentation. With respect to supertask theory it must be recalled that it is an infinitist theory based, as set theory, on the Axiom of Infinity (introduced in Section 3). It originated as a consequence of a seminal discussion about the possibility, or impossibility, of performing an infinite number of actions (tasks) in a finite time interval. [319, 33, 318, 334, 21].

Although the main objective of supertask theory was, and continue to be, the discussion on the actual infinity, its physical implications (including special relativity) have also been discussed in the last years [255, 262, 266, 290, 127, 129, 128, 262, 263, 264, 76, 265, 241, 7, 8, 267, 337, 147, 74, 75, 241, 73, 300]. In short:

> A supertask consists in performing an infinite sequence of actions $\langle a_i \rangle$ within a finite time interval $[t_a, t_b)$, each action a_i being performed at the precise instant t_i of a strictly increasing and convergent sequence of instants $\langle t_i \rangle$ within $[t_a, t_b)$, being t_b the mathematical limit of $\langle t_i \rangle$.

where the elements of $\langle a_i \rangle$ and $\langle t_i \rangle$ are ordered in the same way as the set of the natural numbers in their natural order of precedence: ω-order: 1, 2, 3,.... Notice in this ordering the set exist as a complete totality (Definition 3.2) and each element n has an immediate successor $n+1$ (Peano's Axiom of the Successor, [252, p. 1]), where immediate successor is defined according to:

Definition 3.4 (of Immediate Successor) *All elements of an ordered set A succeeding (preceding) a given element n of A are successors (predecessors) of n in the considered order of A. An element n of an ordered set A is said the immediate successor (predecessor) of another element m of A if n succeeds (precedes) m in the considered ordering of A and no other element of A exists between m and n in that ordering.*

We are now in the appropriate position to analyze the difference between the actual and the potential infinity, Indeed, consider the list L_n of the natural numbers in their natural order of precedence:

$$L_n = 1, 2, 3, \ldots \tag{1}$$

The list L_n can be considered in two different ways:

a) As a complete totality, i.e. as a list in which every element that could be in the list, is in the list (actual infinity).

b) As an unlimited and uncompletable totality (potential infinity).

According to the Hypothesis of the Actual Infinity, the list L_n of the natural numbers in their natural order of precedence 1, 2, 3,...exists as a *complete totality*, i.e as a totality that contains, all at once, all natural numbers. The ellipsis (...) in $1, 2, 3, \ldots$ stands for *all* natural numbers. For all. The word "actual" in *actual infinity* means, therefore, that all elements of an infinite collection as L_n, exist all at once (in the *act*), as a complete totality. In consequence, the list L_n of the natural numbers in their natural order of precedence is considered as a complete totality despite the fact that no last number completes the list. To assume the Hypothesis of the Actual Infinity means, therefore, to assume that it is possible to complete the incompletable, as Aristotle would surely say [15, p. 291]. Or that the incompletable can exist as complete.

To emphasize this sense of completeness, let us consider the task of counting the successive elements of L_n, i.e. the successive natural numbers 1, 2, 3,...in their natural order of precedence. In agreement with the Hypothesis of the Actual Infinity we could count *all* natural numbers in a finite time, for example in an hour, or in a millisecond. The task of counting all natural numbers in a finite time interval, even in less than a second, is an example of supertask:

- *Count each of the successive natural numbers 1, 2, 3...*
  at each of the successive instants t_1, t_2, t_3... of a stric-
  tly increasing sequence of instants $\langle t_i \rangle$ within the finite
  real interval (t_a, t_b), being t_a and t_b any two instants
  such that $t_a < t_b$, and t_b the mathematical limit of the
  sequence $\langle t_i \rangle$. For instance, the classical sequence de-
  fined by:

$$t_n = t_a + (t_b - t_a) \frac{2^n - 1}{2^n} \tag{2}$$

As we will now prove, at t_b all natural numbers would have been counted. All. In effect, let each natural number n of the list L_n be counted at the precise instant t_n of $\langle t_i \rangle$. Being t_b the limit of $\langle t_i \rangle$, t_b is the first instant after all instants of $\langle t_i \rangle$, and all those instants do exist as a complete totality according to the Hypothesis of the Actual Infinity. So, the one to one correspondence f between L_n and $\langle t_i \rangle$ defined by:

$$f(n) = t_n, \ \forall n \in L_n \tag{3}$$

proves that at t_b all natural numbers of the list L_n has been counted. All. The reader can easily imagine why ellipsis and corresponden-ces between sets are the key instruments for demonstrations in infinitist mathematics. Note, on the other hand, that the fact of pairing the elements of two infinite sequences (in our case the one of natural numbers and the other of instants) does not prove both sequences exist as complete totalities. They could also be po-tentially infinite with the same number of elements, a possibility usually ignored in modern infinitist mathematics.

The alternative to the Hypothesis of the Actual Infinity is the Hypothesis of the Potential Infinity, which rejects the existence of *complete* infinite totalities, and then the possibility to count all natural numbers. From this perspective, the natural numbers result from the *endless* process of counting: it is always possible to count a number greater than any given number (Peano's Axiom of the Successor [252, p. 1]). But it is impossible to complete the process of counting all of them, simply because there is not a last natural number to complete the process. So, the complete list of all natural numbers makes no sense, simply because it is incompletable.

The word "potential" in *potential infinity* means, therefore, that the elements of an infinite collection do not exist all at once, but potentially, as possible. The potential infinity is *the unlimited*, as the list L_n of the natural numbers in their natural order of precedence, but only finite collections can be considered as complete totalities, as large as wished but always finite. Similarly, only finite natural numbers can be considered, as large as wished but always finite. For the potential infinite there is not a last natural number (it is always possible to consider a number greater than any previously considered number), but neither is there the complete collection of *all* natural numbers. Contrarily to the actual infinity, the potential infinity assumes the incompletable cannot be completed, cannot exist as complete, precisely because it is not completable.

In short, the actual infinite hypothesis states that the infinite collections are complete totalities, even if no last element completes the collection, as in the case of the ordered list of the natural numbers. On the contrary, the hypothesis of the potential infinite proposes that the infinite collections do not exist as complete totalities, the only complete totalities are the finite totalities, though they can be unlimited in the number of their possible elements. All of which can be summarized in the following definition:

Definition 3.5 (of Actual and Potential Infinity) *An ordered collection of elements is infinite if there is no last (first) element that completes (initiates) it. The collection is actually infinite if it is considered a complete totality, and potentially infinite if it is not considered a complete totality.*

where collection is by set, succession, sequence, list, etc. To be formally precise, the words *set, succession, sequence*, etc. should be replaced by the more general word *collection*. However, for the sake of brevity, it will not be necessary to do so. Therefore, in what follows all of them will be interchangeable with each other, unless otherwise specified.

The potential infinity (the 'improper' or 'non-genuine' infinity as Cantor called it [49, p. 70]) has never deserved the attention of contemporary mathematics. The infinity in Dedekind's Definition 3.3 of infinite set is the actual infinity (see next section). The infinitely many elements of an infinite set exist all at once, as a complete totality. Dedekind's Definition 3.3 is, therefore, based on the violation of the old Euclidean Axiom of the Whole and the Part (the whole is greater than the proper part) [95]. Set theory has been built on that violation.

The hegemony of the actual infinity in contemporary mathematics is absolute. As absolute as the submission of physics to infinitist mathematics. Some authors proceed as if the existence of complete infinite totalities had been formally demonstrated. Obviously, if that were the case we would not need the Axiom of Infinity to legitimize the existence of such infinite totalities. The Hypothesis of the Actual Infinity is just a hypothesis, not a proven fact. And physics should not be subject to infinitist mathematics. In fact, and in agreement with P. Dirac, it should not be subject to any kind of mathematics at all [71, p. VIII]:

> Mathematics is only a tool and one should learn to hold physical ideas in one's mind without reference to the mathematical form.

The three most important "proofs" of the existence of actual infinite totalities (by Bolzano, Dedekind and Cantor) are illustrative of what we could call *naive infinitism*. They also explain why modern infinitist mathematics had finally to establish the existence of actual infinite sets by an arbitrary law, i.e. by means of an arbitrary axiom (the Axiom of Infinity, which is introduced in the next section).

• Bolzano's proof goes as follow (taken from [231, p 112]):

> One truth is the proposition that Plato was Greek. Call this p_1. But then there is another truth p_2, namely the proposition that p_1 is true [But then there is another truth p_3, namely the proposition that p_2 is true]. And so *ad infinitum*. Thus the set of truths is infinite.

But the existence of an endless process (p_1 is true, then p_2 is true, then p_3 is true, then ...) does by no means prove the existence of a final result as a complete totality. At best it proves the existence of an endless (potentially infinite) process. But it does not prove the existence of an actual infinite totality.

• Dedekind's proof is similar (taken from [231, p 113]):

> Given some arbitrary thought s_1, there is a separate thought s_2, namely that s_1 can be object of thought [there is a separate thought s_3, namely that s_2 can be object of thought]. And so ad infinitum. Thus the set of thoughts is infinite.

The above comment on Bolzano proof also applies here. Dedekind gave another proof a little more detailed, albeit with the same formal defect, based on his definition of infinite set [68, p. 115].

• And finally, Cantor's proof: ([133, p 25], [231, p. 117]):

> Each potential infinite presupposes an actual infinity.

or ([47, p. 404] English translation [283, p. 3]):

> ... in truth the potential infinity has only a borrowed reality, insofar as a potentially infinite concept always points towards a logically prior actually infinite concept whose existence it depends on.

But this is an opinion, not a formal proof. It is now clear why the existence of an actual infinite set had to be finally established by law; that is, by means of an axiom.

Let us, finally, state a conventional use of the expressions actual infinite and potential infinity in this and the subsequent chapters of this book. From now on, and for sake of simplicity, the actual infinity will be referred to simply as infinity or actual infinity, while the potential infinity will always be referred to as potential infinity. Or put another way, the word "infinity" will always mean actual infinity, unless it is preceded by the word "potential", in which case it will obviously mean potential infinity. For the sames reasons of simplicity, the world "universe" will always denote the observable universe.

3.2 The infinity of the Axiom of Infinity

Nothing we have been able to observe and measure so far has been infinite. Nor has it been possible to divide anything into an infinite number of parts. On the other hand, and after more than twenty-seven centuries of arguments and discussions, it was not possible to prove (or disprove) the existence of the actual infinities. Infinitism had no choice but to accept that existence in axiomatic terms by means of the Axiom of Infinity. An axiom that simply states the existence of an infinite set:

Axiom 3.1 (of Infinity (ordinary language)):

> *There exists an infinite set.*

Or in abstract, symbolic, terms:

Axiom 3.2 (of Infinity (abstract form)):

$$\exists A : \emptyset \in A \wedge \forall a \in A \ (a \cup \{a\} \in A) \tag{4}$$

that reads: there exists a set A such that $\emptyset$ (the empty set) belongs to A and for every element a in A, the element $a \cup \{a\}$ also belongs to A. Although it is not explicitly declared the type of infinity involved in the set A, it can be easily proved that it is the actual infinity:

Theorem 3.1 (of the Actual Infinity) *The infinity in the Axiom of Infinity can only be the actual infinity.*

Proof: Since potentially infinite sets do not exist as complete totalities, only two subsets with the same number of elements of the same potentially infinite set could be put into a one to one correspondence, and then Dedekind Definition 3.3 is not satisfied, because we would have a one to one correspondence between two proper subsets of a potentially infinite set, in the place of a one to one correspondence between a set and one of its proper subsets. In consequence, the infinity involved in the Axiom 3.1 of Infinity can only be the actual infinity. □

Obviously, an axiom is just an axiom, i.e. a statement that can be accepted or rejected. Some relevant authors as L.E.J. Brouwer, C. Hermite, S. Kleene, J. König, L. Kronecker, H. Poincaré, A. Robinson, L. Wittgenstein, or H. Weyl, among others, rejected the Axiom of Infinity, more or less explicitly. H. Poincaré went so far as to say that (quoted in [231, p. 121], [65, p. 1]):

> infinity is a perverse pathological illness that would one day be cured.

But the vast majority of contemporary mathematicians and physicists do not question the Axiom of Infinity. Indeed, in our days the criticism of the actual infinity is practically non-existent. And infinitism has become a current of thought absolutely hegemonic and quite intolerant of dissent, as if the existence of the actual infinite had been proven. And no, it has not been proven; it has been assumed. And one has the right and the duty to question that assumption, without being insulted and ostracized for it (as is currently the case).

3.3 A short proof of inconsistency

Over the last 30 years, and from different perspectives (set theory, supertask theory, transfinite cardinals, transfinite ordinals, transfinite arithmetics, geometry) I have developed more than forty formal proofs of the inconsistency of the Hypothesis of the Actual Infinity [198]. This section includes one of them, chosen for its brevity and simplicity: the next Theorem 3.5. First, however, it is necessary to consider the following formal elements:

Definition 3.6 (of Inconsistent Set) *A set is inconsistent if a contradiction can be deduced from the number of its elements, or from the number of elements of at least one of its proper subsets.*

Corollary 3.1 (of Inconsistent Sets) *A set with the same number of elements as an inconsistent set, is also inconsistent.*

Proof: It is an immediate consequence of Definition 3.6. □

Definition 3.7 (of Denumerable Set) *A set is denumerable if its cardinal is the smallest infinite cardinal $\aleph_0$ of the infinite set of all natural numbers. An infinite set is non-denumerable if its cardinal is greater than the smallest infinite cardinal $\aleph_0$.*

Definition 3.8 (of ω-Ordered Sets) *A set is ω-ordered if being denumerable, it has a first element, each element has an immediate successor and an immediate predecessor, except the first one which has no predecessor.*

Theorem 3.2 (of Denumerable Sets) *It is always possible to define a one-to-one correspondence between any two denumerable sets.*

Proof: Let A and B be any two denumerable sets. They have the same number of elements: $\aleph_0$ elements (Definition 3.7). So their respective elements can be put into one-to-one correspondence, i.e. each of the different elements of A can be paired with a different and exclusive element of B, so that all elements of A and B result exclusively paired. □

Theorem 3.3 (of non-Denumerable Sets) *Every non - denumerable set has denumerable proper subsets.*

Proof: Let X be any non-denumerable set. Since its cardinal is greater than $\aleph_0$ (Definition 3.7), X contains proper subsets with only $\aleph_0$ elements, all of which are denumerable proper subsets of X (Definition 3.7). □

Theorem 3.4 (of Indexation) *The elements of a denumerable set can be reordered with the same order as the elements of any other denumerable set.*

Proof: Let $A = \{a, b, c, \dots\}$ and $B = \{\alpha, \beta, \dots\}$ be any two denumerable sets. There exists at least one bijection f between the elements of A and B (Theorem 3.2). Consequently, f pairs each element k of A with a unique and exclusive element, say δ, of B, which can be used to exclusively index that element k of A, so that element k can be rewritten as a_δ. Consequently, the elements of the set A can be reordered and rewritten to define the set $A' = \{a_\alpha, a_\beta, a_\gamma, \dots\}$ which has exactly the same elements as A, and ordered in the same way as the elements of B. $\square$

The infinity of infinite sets is the actual infinity, not the potential infinity (Theorem 3.6 of the Axiom of Infinity). This implies the existence of certain infinite sets that are also complete totalities (Definition 3.2). For example the set $\mathbb{N}$ of ALL natural numbers in their natural order of precedence. It is not possible, then, to add new natural numbers to the set $\mathbb{N}$ of natural numbers because it already contains them all. And the same is true of many other numerical or non-numerical sets. For many authors, the existence of these ordered and complete totalities without a last element that completes them (or without a first element that initiates them) is a proven conclusion independent of the Axiom of Infinity. It is not. It is an existence assumed and legitimized by the Axiom of Infinity. Their existence is, therefore, as debatable as the Axiom of Infinity itself. So it is as legitimate to argue about that axiom as it is to argue about the existence of those complete totalities. This fully justifies the following:

Theorem 3.5 (of Denumerable Infinity) *All denumerable sets are inconsistent.*

Proof: Let A be any denumerable set. The set A allows us to define the set A' with the same elements as A but reordered as the set $\mathbb{N}$ of natural numbers in their natural order of precedence: $A' = \{a_1, a_2, a_3, \}$ (Theorem 3.4). The open interval of rational numbers $(0, 1)$ is densely ordered in the natural order of precedence (represented by the symbol $<$) defined by the natural values of the rational numbers. It is also a denumerable set, so there exists a bijection f between A' and $(0, 1)$ (Theorem 3.2). Consequently, $(0, 1)$ can be reordered and rewritten as the set $\mathbb{Q}_{01} = \{q_{a_1}, q_{a_2}, q_{a_3}, \dots\}$, where $q_{a_i} = f(a_i), \forall a_i \in A'$, and the successive elements $q_{a_1}, q_{a_2}, q_{a_3}, \dots$ of $\mathbb{Q}_{01}$ are ordered by the successive natural numbers in their natural order of precedence, and not by their respective values as rational numbers. Let x now be a rational variable defined initially as q_{a_1}. And let the value of x be $<$-compared (i.e., compared according to the values of the rational numbers) with the successive elements of the set $\mathbb{Q}_{01}$, with x being redefined as the compared element q_{a_i} if, and only if, $q_{a_i} < x$.

For short, let us call comparison* this $<$-comparison and redefinition of x if, and only if, the value of the compared element is smaller than the current value of x. It is immediate to prove that for each natural number v it is possible to perform the first v comparisons* of x with the first v successive elements of $\mathbb{Q}_{01}$. Indeed, if it were not possible, there would be at least one natural number $n \leq v$ such that x could not be compared* with q_{a_n}, which is impossible because q_{a_n} is a rational number of $\mathbb{Q}_{01}$ that can be compared* with the current value of x, which is also a rational number. Once all possible comparisons* of x with the successive elements $q_{a_1}, q_{a_2}, q_{a_3}, \ldots$ of $\mathbb{Q}_{01}$ have been made, the current value of x, whatever it may be, could only be the smallest rational number of that set. Indeed, if once performed all possible comparisons* of x with the successive elements of $\mathbb{Q}_{01}$ the current value of x were not the smallest rational number of $\mathbb{Q}_{01}$, there would be at least one element q_{a_n} in $\mathbb{Q}_{01}$ such that $q_{a_n} < x$. But that is impossible because n is a natural number; the first n comparisons* have been carried out; and therefore x was compared* with q_{a_n} and redefined as q_{a_n}; and in all subsequent comparisons*, x could only be redefined with values smaller than q_{a_n}. Therefore, it is impossible for $q_{a_n} < x$. But, on the other hand, it is also immediate to prove that once all possible comparisons* of x with the successive elements of $\mathbb{Q}_{01}$ have been made, the current value of x is not the smallest rational number of that set: every element of the infinite set $\{x/2, x/3, x/4 \ldots\}$ is an element of $\mathbb{Q}_{01}$ smaller than x. This contradiction proves that the set A', defined exclusively with the elements of A, is inconsistent. Therefore A' and A are inconsistent (Definition 3.6). And A being any denumerable set, it must be concluded that all denumerable sets are inconsistent. □

Although the consistency of a mathematical proof of infinite steps is universally accepted without the need to perform all of its infinite steps, the theory of supertasks considers the possibility of performing them in finite time. In the case of the above successive comparisons* of x with each successive q_{ai} would be performed at each successive instant t_i of a strictly increasing and convergent sequence $\langle t_i \rangle$ of instants within the finite time interval (t_a, t_b), whose limit is t_b. The instant t_b is the first instant after all instants of $\langle t_i \rangle$, and therefore the first instant after having performed all possible comparisons* of x with the successive elements of Q_{01}. At the instant t_b the rational variable x will still be a rational variable with a certain value, whatever it is; and not, for example, an elephant (in which case anything could be proved). The problem is that the value of x at the instant t_b is and is not the least rational of Q_{01}.

Corollary 3.2 (of Inconsistent ω-Order) *ω-ordered sets are inconsistent.*

Proof: Since ω-ordered sets are also denumerable sets (Definition 3.8), they are inconsistent (Theorem 3.5). $\square$

From the previous theorems and corollaries, we can immediately deduce, among many others, the following results:

3.4 The axiom of infinity is inconsistent

The above Theorem 3.5 proves the inconsistency of any denumerable set. It is then immediate to prove the following results:

Theorem 3.6 (of the Axiom of Infinity) *The Axiom of Infinity is inconsistent.*

Proof: Let us write the set A defined in Axiom 3.2:

$$\exists A : (\emptyset \in A \wedge \forall a \in A \ (a \cup \{a\} \in A)) \tag{5}$$

as:

$$A = \{a, s_1(a), s_2(a), s_3(a), \dots\} \tag{6}$$

where:

$$s_1(a) = a \cup \{a\} \tag{7}$$
$$s_2(a) = s_1(a) \cup \{s_1(a)\} \tag{8}$$
$$s_3(a) = s_2(a) \cup \{s_2(a)\}\} \tag{9}$$
$$s_4(a) = s_3(a) \cup \{s_3(a)\}\} \tag{10}$$
$$s_5(a) = s_4(a) \cup \{s_4(a)\}\} \tag{11}$$

$$\cdots$$

Consider now the set $\mathbb{N}$ of the natural numbers, which is denumerable, and the set A defined by (6), which is the set whose existence claims the Axiom of Infinity. The one to one correspondence

f between the denumerable set $\mathbb{N}$ and A defined according to:

$$f(n) = s_n(a), \ \forall n \in \mathbb{N} \tag{12}$$

proves that A is also an inconsistent set (Theorem 3.5 and Corollary 3.1). $\square$

And from Theorems 3.1 and 3.6 it immediately follows the next three corollaries:

Corollary 3.3 (of the Inconsistent Infinity) *The actual infinity is inconsistent.*

Proof: It is an immediate consequence of Theorems 3.1 and 3.6. $\square$

Corollary 3.4 (of the Actual Infinite Sets) *All actual infinite sets are inconsistent.*

Proof: It is an immediate consequence of Theorems 3.1 and 3.6. $\square$

Corollary 3.5 (of Consistent Collections) *A set can be either a finite complete totality or a potentially infinite and uncompletable totality. Otherwise it is inconsistent.*

Proof: It is an immediate consequence of Definition 3.5 and Corollary 3.4. $\square$

Let us now recall the following definition:

Definition 3.9 (of Densely Ordered Sets) *If no element of a strictly ordered set has an immediate predecessor nor an immediate successor, the set is said to be densely ordered or to define a continuum.*

We can now prove the following:

Theorem 3.7 (of the Inconsistent Dense Order) *Densely orde - red sets are inconsistent.*

Proof: Let X be a densely ordered set. Suppose X is finite. It will have a finite number of elements, say n. Let x_1 and x_2 be two elements of X such that x_2 is a successor of x_1. Since x_2 cannot be the immediate successor of x_1, there will exist between x_1 and x_2 at least one other successor x_3 of x_1. Since x_3 cannot be the immediate successor of x_1, there will exist between x_1 and x_3 at least one other successor x_4 of x_1. By repeating this argument

$n-2$ times we will arrive at a successor x_{n-2} of x_1 that would have to be its immediate successor, which is impossible. Therefore, X cannot be finite. And being infinite it is inconsistent (Corollary 3.4). $\square$

Corollary 3.6 (of the Inconsistent $\mathbb{Q}$ and $\mathbb{R}$) *When considered as complete infinite totalities, the set $\mathbb{Q}$ of the rational numbers and the set $\mathbb{R}$ of the real numbers are both inconsistent.*

Proof: It is an immediate consequence of Corollary 3.4, and also of Theorem 3.7, because they are densely ordered sets. $\square$

Theorem 3.8 (of the Inconsistent Continuum) *The spacetime continuum is inconsistent.*

Proof: The spacetime continuum is the Cartesian product (cross product) of sets $\mathbb{R}^4 = \mathbb{R} \times \mathbb{R} \times \mathbb{R} \times \mathbb{R}$, each of whose factors is the set $\mathbb{R}$ of real numbers. Consequently it is an inconsistent set (Corollaries 3.6 and 3.1). $\square$

The above results on the inconsistency of the infinite sets, including the inconsistency of the continuum and of densely ordered sets, will change everything. So deconstructing the arguments that follow here and in the subsequent articles of this series of articles, will involve proving the falsity of Theorem 3.5 (and the falsity of each of the more than 40 independent proofs included in [198]).

Let us now consider the following:

Definition 3.10 (of Discrete Sets) *A set is discrete if it has a first element, a last element and each of its elements (except the first one) has an immediate predecessor and (except the last one) an immediate successor.*

And then, let finally prove the following

Theorem 3.9 (of Discrete Sets) *All discrete sets are finite.*

Proof: Let A be any discrete set:

$$A = \{a, s_1(a), s_2(a), s_3(a) \dots s_v(a)\} \tag{13}$$

where $s_1(a)$ is the immediate successor of a; $s_2(a)$ the immediate successor of $s_1(a)$; $s_3(a)$ the immediate successor of $s_2(a)$; and so on. If an element $s_n(a)$ has a finite number n of predecessors, then its immediate successor $s_{n+1}(a)$ has also a finite number $n+1$ of predecessors: all n predecessors of $s_n(a)$ plus $s_n(a)$. And since the element $s_1(a)$ has a finite number of predecessors, just 1 predecessor, the element a, we can inductively conclude that all elements of A, including its last element, have a finite number of predecessors. Therefore, A has a finite number of elements. $\square$

Theorem 3.10 (of the Strictly Ordered Sets) *Every strictly or - dered set is discrete.*

Proof: Let a be any element of any strictly ordered set A, and suppose A has not a last element. Since a is not the last element of A, there exist successors of a in A. Let us consider one such successors and denote it by a_1. For the same reasons as in the case of a, we can consider and denote by a_2 any successor of a_1 in A. For the same reasons as in the case of a_1, we can consider and denote by a_3 any successor of a_2 in A. For the same reasons as in the case of a_2, we can consider and denote by a_4 any successor of a_3 in A. We would thus have a sequence of successors of a: a_1, a_2, a_3, $a_4 \ldots$ in which there is not a last element. The bijection f between A and the ω-ordered set $\mathbb{N}$ defined by $f(a_i) = i$ proves that A, like $\mathbb{N}$, would be infinite, and therefore inconsistent (Corollary 3.4). Consequently, A has a last element. Exactly the same argument now referring to the predecessors of a, proves also that A has a first element. Let a now be any element of A other than the last element of A. Suppose that a has not an immediate successor. Let a_1 be any successor of a. Since a_1 is not the immediate successor of a there will exist another successor a_2 of a between a and a_1. Since a_2 is not the immediate successor of a there will exist another successor a_3 of a between a and a_2. The same argument above shows that the sequence of successors a_1, a_2, $a_3 \ldots$ of a is inconsistent. Therefore a has an immediate successor. The same argument now referring to any element b different from the first element of A proves that b has an immediate predecessor. Consequently, A is discrete (Definition 3.10). $\square$

Theorem 3.11 (of Discrete Sets) *Every set is either discrete or discretely orderable.*

Proof: Let A be any set. If it is strictly ordered, it is a discrete set (Theorem 3.10). If it is unordered and consistent, it will have a finite number n of elements. By a bijection f, each of its elements can be paired with a different natural number of the set $\mathbb{N}_n$ of the first n natural numbers in their natural order of precedence. The set A^* defined by f^{-1}:

$$A^* = \{f^{-1}(1), f^{-1}(2) \ldots f^{-1}(n)\} \tag{14}$$

is an ordered version of A, and therefore a discrete version of A (Theorem 3.10). $\square$

As noted above, more than forty other different and independent arguments included in [198] reach the same conclusion about the inconsistency of the Hypothesis of the Actual Infinity subsumed in the Axiom of Infinity. This infinity is what Aristotle would surely call infinite by addition. In the next paper 4, it will be proved the inconsistency of the other Aristotelian infinitude: the infinite by division, which was the type of infinite involved in the formalized version of Zeno's Dichotomies I and II [45, 46, 328, 329, 290, 154, 333, 62, 220, 125, 126, 342, 127, 129, 128, 222, 221, 207, 208, 246, 6, 266, 290, 154, 300].

Physical models and theories work reasonably well (even very well) until the infinities appear. But physicists do not usually concern themselves with the formal consistency of the infinitist mathematics that they use in all their models and theories. Nor do they concern themselves with another problem essential to a consistent explanation of the physical world: the problem of the infinite regress (of proofs, definitions, and causes). As will be seen throughout this series of articles, it is possible to modify the infinitist models and theories used in physics by finitist and discrete versions in such a way that they remain compatible with all the accumulated empirical knowledge about the physical world. And, at the same time, they are much simpler, more physical and less extravagant than their infinitist counterparts.

3.5 Conclusion

If any one of the more than forty proofs of the inconsistent nature of the actual infinity given in [198] is right, then the Hypothesis of the Actual Infinity is inconsistent. One of those arguments has been reproduced here so that the reader can directly evaluate the possibility that, in fact, the Hypothesis of the Actual Infinity, and then the Axiom of Infinity were inconsistent. If so, we might draw our first two cosmological conclusions:

Theorem 3.12 (of the Finite Universe) *A consistent universe cannot contains an actual infinite number of physical objects.*

Proof: It is an immediate consequence of Corollary 3.4. $\square$

According to the Standard Model there exists a finite number of different elementary particles (six quarks, six leptons and five bosons), each with a different finite mass. Therefore, the following is also true:

Corollary 3.7 (of the Finite Mass-Energy) *The mass and the energy of the observable universe cannot be actually infinite.*

Proof: It is an immediate consequence of the Standard Model, Theorem 3.12 and the mass-energy relation. □

Paper 4. Discrete versus continuous

Abstract. This paper confronts discreteness with continuity, and applied the confrontation to physical magnitudes, most of which are already defined as discrete magnitudes (quantum magnitudes). After recalling the pre-Socratic concept of the continuous (initially made of extensive points) and the modern spacetime continuum (which was proved to be inconsistent in article 3 of this series of articles), the inconsistent nature of the real numbers with infinitely many decimals is demonstrated when those sequences of decimals are considered as complete totalities. The inconsistency of the infinite division of space and time is then proved, a result of the greatest importance from the discrete perspective of space and time that will be developed in the subsequent articles of this series of articles. Finally, the lack of immediate successiveness (adjacency) in the continuum is used to introduce the problem of change, a pre-Socratic question still unsolved, not even by physics, the science of change.

Keywords: continuous, extensive points, spacetime continuum, non-computable numbers, numbers with infinitely many decimals, continuous magnitudes, discrete magnitudes, Planck constants, infinite division, immediate successiveness, adjacency, problem of change.

4.1 The problem of the continuous

Although related to the modern spacetime continuum (see next Section 4.2), the problem of the continuous has a Pythagorean origin [213]. In my opinion, its importance in the history of science has not been sufficiently appreciated. The firsts Pythagorean believe in the existence of indivisible geometrical points with an

45

extent δ greater than zero, consequently they believed that all lengths would have to be commensurable: the ratio between any two of these lengths, say L_1 and L_2, would be a ratio between two natural numbers [213, pp. 11-16]:

$$L_1 = n_1\delta; \; L_2 = n_2\delta \tag{1}$$

$$\frac{L_1}{L_2} = \frac{n_1\delta}{n_2\delta} = \frac{n_1}{n_2} \tag{2}$$

Somewhat later, the Pythagorean discovered the existence of non-commensurable lengths: the length of the diagonal L_d of a square with the length of its side. For example, if the length of the side is 9δ, we would have:

$$L_d = \sqrt{9^2\delta^2 + 9^2\delta^2} \tag{3}$$

$$= 9\delta\sqrt{2} \tag{4}$$

$$\frac{L_d}{L_s} = \frac{9\delta\sqrt{2}}{9\delta} = \sqrt{2} \tag{5}$$

Unfortunately, they did not consider the possibility of a discrete

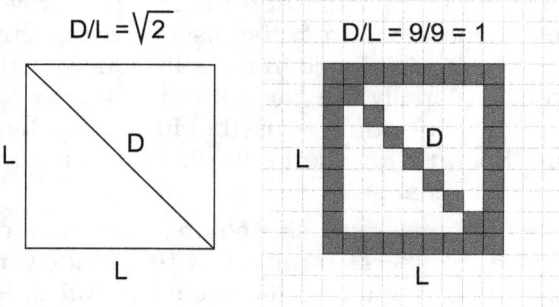

Figure 4.1 – Left: In continuous geometry the diagonal D and the side L of a square are not commensurable. Right: In discrete geometry the diagonal D and the side L of a square are commensurable.

arithmetic, for instance:

$$L_d = \lfloor \sqrt{9^2\delta^2 + 9^2\delta^2} \rfloor \tag{6}$$

$$L_d = \delta \lfloor \sqrt{9^2 + 9^2} \rfloor \tag{7}$$

$$= 9\delta \lfloor \sqrt{2} \rfloor \tag{8}$$

$$= 9\delta \tag{9}$$

$$\frac{L_d}{L_s} = \frac{9\delta}{9\delta} = 1 \tag{10}$$

where $\lfloor x \rfloor$ stands for the integer part of x. As we will see in paper 14, equations (6)-(10) represent the discrete version of Pythagoras theorem.

A second exception on the discrete nature of space and time was the theory of discrete space-time that developed in the IX-X centuries within the Arabic philosophical-theological school of thought known as Kalam (or Kalām) [165, p. 62-68]. According yo this theory even motion would be discrete, discontinuous.

Perhaps due to the enormous influence of our sensory perception of the physical world as a continuous scenario of space and time (see Section 4.7), this type of discontinuous arithmetic is yet to be developed in formal and universal terms. In any case, one of the consequences of the Pythagorean discovery of incommensurable lengths was the abandonment of the extensive points in favor of the non-extensive points, which are the same points that we still use today in all continuous geometries. But the discussions about the continuous (and in general about infinity) lasted until the beginning of the 20th century, when the existence of an actual infinite set was axiomatically proposed and accepted in the nascent set theory. Since then, the hegemony of infinitist mathematics (and geometry) has been as absolute as the submission of physics to infinitist mathematics. As is often the case with hegemonic positions (even of thought), in addition to being hegemonic they have been, and continue to be, quite hostile to dissidence, which has been ostracized for more than a century. The power of the academy is not different from other more or less absolutist human powers.

4.2 **The spacetime continuum**

The word *"continuum"* is a rather polysemous term in mathematics and physics. It can be used, for example, as:

- The set of the real numbers (Harper Collins Dictionary of Mathematics [40, p. 118]).

- A continuous distribution of matter (Harper Collins Dictionary of Mathematics [40, p. 118]).

- A system of axes that form a frame of reference (Oxford Dictionary of Physics [64, p. 94]).

- The set of points on a line (linear continuum) (Oxford Dictionary of Philosophy [34, p. t8]).

- A compact connected metric space (Wolfram MathWorld [336]).

- Said of a magnitude: that takes values that are not separate from each other (DRAE).

The expression "power of the continuum" is less polysemous: it is the cardinal $2^{\aleph_o}$ of the set $\mathbb{R}$ of the real numbers, or of any of its non-denumerable subsets. Although the word "power" in "power set" also means: the set of all subsets of a given set. And the power of the continuum is also the cardinal of the power set of the set $\mathbb{N}$ of the natural numbers, i.e. the cardinal of the set of all subsets of the set $\mathbb{N}$ of the natural numbers when considered as a complete totality (Definition 3.2), which is the natural consideration in infinitist mathematics. For any given set A, its cardinal number is usually written $|A|$, and its power set $P(A)$. So, according to the above definitions, we can write the well known equalities:

$$|\mathbb{N}| = |\mathbb{Q}| = \aleph_o \qquad (11)$$

$$|\mathbb{R}| = |P(\mathbb{N})| = |P(\mathbb{Q})| = 2^{\aleph_o} \qquad (12)$$

where $\mathbb{Q}$ and $\mathbb{R}$ are respectively the set of (all) rational numbers and the set of (all) real numbers. In physics, the continuum is almost always associated with spacetime, the relativistic assumed four - dimensional manifold of space and time. In fact, relativistic spacetime is a four-dimensional continuum of spatial points and temporal instants modeled by the set $\mathbb{R}^4$ of all real 4-tuples (x, y, z, t).

One of the most picturesque properties of the infinitist spacetime continuum is that any linear segment, for instance of a Planck's length:

$$l_p =\approx 0.0000000000000000000000000000000016 \text{ millimeters}$$

has the same number of points as the entire observable three - dimensional universe. Or that any interval of time, for example the duration of Planck time:

$$t_p =\approx 0.00000000000000000000000000000000000053 \text{ seconds.}$$

has the same number of instants as the entire history of the observable universe (over 13800 million years). As is well known, this is the so called Dimension Problem proved by Cantor[1] [16, 69, 301, 333, 122, 65, 50, 57]. It is only necessary to define the corresponding one-to-one correspondence between both sets and make an appropriate use of ellipsis (the magic wands of infinitism) to prove it. This has little discussed consequences in physics: for example, infinitist physics assumes that a segment of Planck length contains as many point entities (point charges, point masses, virtual point particles etc) as the whole three dimensional observable universe.

The concept of point, on the other hand, is a primitive concept; an undefined concept of which we do not even have an intuitive notion (as is the case, for example, of the primitive concept of set). And the same can be said of the concept of instant, "*the points of time*". The mark we make on a sheet of paper with the tip of a pencil, or on a blackboard with chalk, is not a point but a conventional graphic mark drawn for representational purposes, a graphic mark that, by the way, contains the same number of points as the whole universe. I have the impression that physicists are not interested in these logical problems related to the actual infinity, although they are anything but irrelevant to physical models and theories. In papers 9 and 10 of this series we will discuss respectively on discrete space and discrete time. And in paper 11 it will be *formally* proved that points can have neither size nor shape, which, as will be discussed there, has very important physical consequences.

The most significant use of the continuum in physics is in the spacetime continuum, a tetra-dimensional continuum of spatial points and temporal instants modeled, as noted above, by the set $\mathbb{R}^4$, the set of all 4-tuples (x, y, z, t) of real numbers. The spacetime continuum constitutes the fundamental core of special relativity, which at the end is a theory of the spacetime continuum [317, 227]. The interest in the geometry of space and in the nature of time, including the debates on their infinite divisibility, has interested philosophers at least since the 14th century [204], and continues to be in the center of a few number of modern philosophical debates [269, 238, 216, 217, 143], now including the (still irrelevant) alternative of a discrete and finite space and time.

---

[1]It is also well known Cantor's reaction of not publishing any more in the magazine The Journal of Crelle that rejected the referred proof [69], the same reaction of Einstein with the magazine The Physical Review that in 1936 dared to review one of his articles [173, 245]

Some pioneering authors were interested in discrete spaces in the first half of the 20th century [31], [61]. W. Heisenberg, for instance, conceived the idea of space as a kind of crystal lattice made up of tiny cells of the size of elementary particles, although he did not develop the idea in the end. Things have started to change, especially in the last two decades [108, 225, 176, 112, 22, 279, 23, 17, 274]. A growing number of physicists now suspect that Planck length and Planck time define a kind of granularity of space and time that could be an effective alternative to the infinitist continuum; an alternative that could be experimentally tested [236, 63, 109, 200, 54]. The problem is that, although the discrete nature of spacetime has been proposed in different areas of physics [165, 124, 325, 100, 303, 304, 312, 20, 203, 211, 9, etc.], the proposals have been invariably developed in the framework of infinitist mathematics. There can be no greater blunder. To briefly examine the involvement of points and instants in physical phenomena, we will end this section by recalling two well-known physical phenomena involving the points and instants that supposedly form the spacetime continuum: the diffraction of light through a narrow slit (Fraunhofer diffraction), and the spacetime propagation of a field perturbation. In relation to the first phenomenon, whose current explanation is also well know, you can read things like:

- Each *point* of the wavefront become a secondary source of waves, emitting new waves, called diffracted waves ...

- Each *point* of the slit become a secondary source of waves ...

- Each photon at a *point* ...

- etc

But there is a problem with this classical explanation (a problem, by the way, representative of the lack of rigor in the use of language): any slit has the same number of points as any other, which is also the number of points in the whole observable universe: $2^{\aleph_o}$ points. Therefore, for a given source of light of a given wavelength and a given distance from the slit to the viewing screen, the corresponding diffraction pattern cannot be explained in terms of points: since all slits have the same geometry and the same number of points, they would all produce the same diffraction pattern, whatever the slit width. And this is not the case: the pattern also depends on the width of the slit.

We could consider the alternative that some points are involved in the formation of the pattern and some others are not. But which ones are involved and which ones are not? We must conclude that the diffraction of light through a single Fraunhofer slit is not satisfactorily explained by infinitist mathematics, in this case by the geometry of the infinitist continuum. And the reason is just the supposed continuum of points space is made of. Indeed, this problem does not arise within the framework of CALMs (Cellular Automata Like Model) introduced in paper 6 of this series of papers.

Even more dramatic is the situation with the propagation of a field (e.g. electromagnetic) perturbation through the spacetime continuum. Here we will inevitably find erroneous expressions such as: adjacent points, contiguous points, successive points, next instants, successive instant etc. all of them impossible: in the spacetime continuum no point (instant) has an immediate successor, a next, adjacent, contiguous... point (instant) as, for example, 4 has its immediate successor 5 in the case of natural numbers. The points of a continuum do not touch each other. It is only possible to jump from one point (instant) to another through a non-numerable infinity of points (instants): again $2^{\aleph_o}$ points (instants). So, in the end, only discontinuous and incomplete descriptions of this propagation through the spacetime continuum can be made, incomplete because it is not possible to know what will be the next point (instant) affected; and it will not be possible because there is no point (instant) immediately succeeding (adjacent to) a given point (instant). A real drama ignored by the theoretical infinitist physics (although it is not aware of its infinitism).

4.3 Finite but non-computable natural numbers

This section proposes to the reader an exercise of imagination and at the same time of humility. It proposes an objective reflection on the incommensurable size of most of the natural numbers included in the list of natural numbers in their natural order of precedence, when that list is considered as a complete totality (Definition 3.2). All these natural numbers are finite, and each of them is just one unit greater than its immediate predecessor, except the first of them: the number 1 which has no predecessor and is defined just by one of such unities. It will be worth trying to imagine the greatness of the vast majority of these finite numbers. I say try to imagine because, as we will see here, it is not even possible to imagine them.

Well, at the end of the 19th and beginning of the 20th century we ended up assuming that all these numbers (all!) exist in the act, as a complete totality (Hypothesis of the Actual Infinity subsumed in the Axiom of Infinity). And that there is another number greater than all of them: $\aleph_o$, the smallest infinite (cardinal) number greater than all finite natural numbers, which is also the total number of natural numbers (the cardinal of the set $\mathbb{N}$ of natural numbers). Or in other words, although there is not a last natural number completing the ordered list of natural numbers, we assume there is a precise number of numbers in that list, and that this number is the least (infinite) number greater than all numbers in that list. This is the foundational core of infinitist mathematics.

Indeed, current infinitist mathematics assumes the existence of the set $\mathbb{N}$ of the natural numbers in their natural order of precedence $\{1, 2, 3,\dots\}$ as a complete totality with just $\aleph_o$ elements, each one unit greater than its immediate predecessor, except the first, and all of them finite. Current infinitist mathematics assumes, therefore, that once defined the least natural number, the number 1, as one unit, infinitely many other natural numbers (all of them finite) can be successively defined simply by adding one unit to the last defined natural number (Peano's Axioms [252, p. 1]). And, above all, that the list of natural numbers so defined exists as a complete totality that contains in the act all! natural numbers (Axiom of Infinity).

So, after adding infinitely many successive units to one initial unit we do not reach a number with infinitely many units, but infinitely many finite natural numbers; each with finitely many units and with one unit more than its immediate predecessor. This is quite conflicting, and the reason for the conflict is the assumption that the ordered list of the natural numbers exists as a complete totality despite the fact that no last natural number completes the list (see [191] or [198, Chapter 31]). But not only conflicting, this assumption is also inconsistent as is proved in paper 3 of this series of articles, and in other more than forty different ways in [198].

Two well known attributes of the natural numbers are their immediate successiveness (adjacency) and their discreteness: except the first one, each natural number n has an immediate predecessor $n - 1$ and an immediate successor $n + 1$, so that there is no other natural number between $n - 1$ and n, nor between n and $n + 1$. In addition, and being all of them finite, between any two natural numbers m and n (being $m < n$) only a finite number $n - m$ of natural numbers do exist. That said, it is worth considering the gigantic size that successive finite natural numbers can reach. It is an unusual but very convenient exercise to have a comparative reference on the size of certain natural numbers and on the number of decimals of the real numbers that are discussed in the next section.

Indeed, it is possible to define in precise arithmetic terms natural numbers that written in standard text (e.g., 5 millimeter per digit) would occupy a length millions of times greater than the diameter of the visible universe. And they are not anodyne sequences of zeros, but precise sequences of different digits (for example, of the decimal numbering system). This is the case of the expo-factorial numbers (in symbols $n^!$, note the factorial symbol ! is here an exponent, a power) and, specially, the case of n-expo-factorial numbers (in symbols $n^{!n}$) [198, Chapter 14]:

> The grandeur of, for example, $9^{!9}$ (9-expofactorial of 9) is far beyond human imagination. Three standard arithmetic symbols, just *9, !, 9*, is all we need to define a finite number so large that the standard writing of its precise sequence of figures would surely be a string of numerals of a length millions of times greater than the diameter of the observable universe. If we use the hexadecimal numeral system, $F^{!F}$ would be inconceivable greater.

It is a good exercise for our imagination (and humility) to imagine a finite natural number that would take us millions of years to go through all its digits, moving at 300000 kilometers per second. Not to say the impossibility to represent all those digits in material terms taking into account that the estimated number of elementary particles of ordinary matter in the observable universe is only in the order of 10^{80} [327, 237]. Obviously, numbers as $9^{!9}$ are finite but uncomputable. They can only be defined theoretically.

Imagine now we place a zero and a decimal point before the first digit of $9^{!9}$. We would have a precise rational number $q_{!9}$ with a finite number $9^{!9}$ of decimals, being its sequence of decimals the same as the sequence of digits of $9^{!9}$, which, uncomputable as it may be, will be a precise sequence of digits, for example (the sequence of digits is invented):

$$q_{!9} = 0.3412983247520983\ldots908734 \qquad (13)$$

Now consider any physical magnitude M whose mathematical definition includes an irrational number as π, e, $\sqrt{2}$ etc. To measure

M with a total precision would require to know the precise ω-ordered sequence of all its decimals, which contains $\aleph_o$ decimals, and then infinitely many decimals more than q_{19}. To measure a physical magnitude as M with a precision of, say 30 decimals, would qualify as an extraordinary success (and in my opinion it would be an extraordinary success). But an extraordinary success only within the finitist scenario, because from the infinitist point of view, 30 decimals are absolutely insignificant compared with its total number of decimals, which exist as a complete totality of $\aleph_o$ decimals. Is it appropriate to talk about the extremely high degree of precision of a measurement with 30 decimal places if the actual number of decimal places is, for instance, 9^{19}, which is infinitely less than $\aleph_o$?

4.4 Numbers with infinitely many decimals

From the infinitist perspective, the infinitely many decimals of a real number with an infinite decimal expansion, as for example π, do exist as a complete and ω-ordered totality, which means that it is a mind-independent entity because our mind cannot embrace the actual infinity (we cannot even imagine a number with a finite number of 9^{19} decimals). But as will be shown below, numbers with an infinite number of decimals are inconsistent.

Let a be the decimal expansion of any real number with an infinite decimal expansion:

$$a = .d_1 d_2 d_3 \cdots = d_1 \times 10^{-1} + d_2 \times 10^{-2} + d_3 \times 10^{-3} + \ldots \qquad (14)$$

and let A be the set of all its successive decimals:

$$A = \{d_1, d_2, d_3, \ldots\} \qquad (15)$$

It is immediate to define a one to one correspondence f between A and the set of natural numbers $\mathbb{N}$:

$$f(d_i) = i, \ \forall d_i \in A \qquad (16)$$

Among dozens of others [198], Theorem 3.5 of the denumerable infinity (see paper 3 of this series) proves the set $\mathbb{N}$ is inconsistent when considered as a complete totality. In consequence, the same must apply to the set A and to the infinite decimal expansion of the above infinite decimal expansion a when considered as a complete totality. Thus, and being a any infinite decimal expansion, the above argument proves the following:

Theorem 4.1 (of Decimal Expansions) *All infinite decimal expansions are inconsistent.*

From the point of view of the potential infinity, things are very different: from this perspective a real number is not a mind independent entity formed by a complete ω-ordered sequence of decimals that exist all at once and by themselves. From this perspective, the irrational numbers result from endless process of calculation that cannot be replaced with a division between two integers, although at each stage of the calculation the number coincides with a rational number with a finite number of decimals. In this sense, the irrational numbers are also definable as (potentially infinite) sequences of rational numbers, and therefore as sequences of ratios between two integer numbers.

In the case of rational numbers, they are defined by a division between two integers, a division that may or may not have an end. In turn, integers would result from the endless process of counting, i.e. from the endless process of adding (or subtracting) successive units to one initial unit. Obviously, the existence of endless processes of counting and calculation does not necessarily mean the existence of their corresponding finished results as complete totalities, as is assumed from the infinitist perspective. It is time to recall the famous quote by Leopold Kronecker (collected in numerous texts, for example in [306, p. 117]):

> God has created the natural numbers, the rest is man's work.

And the commentary appearing on a web page of a certain University:

> He [Kronecker] opposed the work of his student Georg Cantor on infinity, considering that it lacked rigor. How wrong he was!

The author of the above commentary should be reminded that Cantor did not prove the existence of infinities (see paper 3 of this series). The existence of infinities had finally to be established by an arbitrary law (Axiom of Infinity). We shall see who is wrong. It has gone so far in mathematical infinitism, and physics has relied so much on infinitist mathematics, that the consequences of its inconsistency could be the greatest in the history of science. So, check out Theorem 3.5 in paper 3 in this series, or the more than 40 arguments contained in [198].

Finally, a curiosity: if we remove the decimal point from (14) we get:

$$a' = d_1 d_2 d_3 \ldots \tag{17}$$

But which kind of number is a'? It cannot be a natural number because natural numbers are finite and each of them (except the number 1) is greater than the number of its digits (in the decimal numbering system), and since a' has an infinite number of digits it cannot be a natural number (otherwise it would be an infinite natural number). And then, is it a member of the sequence of powers (18), or of the sequence of alephs (19)?

$$\aleph_o, \ 2^{\aleph_o}, \ 2^{2^{\aleph_o}} \ldots \tag{18}$$

$$\aleph_o, \ \aleph_1, \ \aleph_2, \ldots \tag{19}$$

or a new kind of infinite number? In any case it would be as inconsistent as any other infinite number.

4.5 Discrete and continuous magnitudes

The main difference between a discrete magnitude and a continuous magnitude is that in a discrete magnitude there is always an indivisible minimum value; and in a continuous magnitude there is not. A significant consequence is that the set V of all possible values of a discrete magnitude is completely determined by its particular indivisible minimum μ and the successive natural numbers:

$$V = \{\mu, 2\mu, 3\mu, 4\mu, \ldots\} \tag{20}$$

Therefore, each value $n\mu$ of the sequence of successive values of a discrete magnitude has an immediate successor $(n+1)\mu$, and an immediate predecessor $((n-1)\mu$ if $n > 1$; and no other value exist between any value and its immediate successor, or its immediate predecessor, if any. For this reason, the sequence of all possible values of a discrete magnitude is said to have immediate successiveness, adjacency (as the sequence of natural numbers in their natural order of precedence). Furthermore, since the indivisible minimum must be finite (Corollary 3.3), and can only have a finite number of decimals (Theorem 4.1), we can conclude that the sequence of the possible values of any discrete magnitude has immediate successiveness, adjacency (each value has an immediate successor, except the last one) and each of its members is finite and have a finite number of decimals, if any.

Things are quite different with continuous magnitudes because they are densely ordered: between any two values of a continuous magnitude infinitely many other values do exist; and between any two of these infinitely many values, others infinitely many values do exist; and between any two of these infinitely many values, others infinitely many values do exist; and so on and on. Therefore, a sequence of continuous magnitudes has not immediate successiveness (adjacency): between any two values of this continuous sequence of values there is always an infinite number of different values.

The definition of some physical constants involves irrational numbers, as is the case of π in the Planck constants: Planck length, Planck time, Planck mass, Planck energy and Planck charge. So, all of them are irrational numbers with an ω-ordered sequence of decimals, i.e. with infinitely many decimals. From the infinitist perspective all these constants are irrational numbers with an infinite number of decimals. And according to Theorem 4.1 on numbers with infinitely many decimals, all them would be inconsistent when the infinite sequence of their corresponding decimals are considered as complete totalities. So, each of these constants should have a precise finite sequence of decimals, if they are consistent constants. Finally, and with respect to the variable magnitudes, the following is verified:

Corollary 4.1 (of Discrete Values) *The number of all possible values of a variable magnitude is finite, and all of them can be arranged in a discrete set with a minimum and a maximum value.*

Proof: It is an immediate consequence of Corollary 3.4 and Theorems 3.10 and 3.11. □

4.6 Inconsistency of the actual infinite divisions

In the third article of this series it was proved the inconsistency of the ω-ordered sets, and then the inconsistency of any other infinite collection that contains ω-ordered subcollections, as is the case of the continuum. This result suffices to prove the inconsistency of the infinite division of space (or time) because any such division is defined by an infinite ordinal $\alpha \geq \omega$ and an α-ordered sequence of points (instants) [198]. In this section, other independent proofs will be given that confirm the inconsistency of the infinite division of an interval of space or time (in fact, of any finite interval or real numbers). First, the inconsistency of dividing a finite interval of real numbers into an infinite number of equal parts will be demonstrated. And secondly, this same inconsistency will be demonstrated for an infinite number of parts of decreasing size.

Theorem 4.2 (of the Finite Divisions) *Any finite interval of real numbers divided into parts of equal length, except at most the last part if any, can only have a finite number of parts.*

Proof: Let (a, b) be any open finite real interval (the argument also applies to closed and semi-closed real intervals). To divide (a, b) into a certain number of parts means to define a number of adjacent and disjoints sub-intervals:

$$(a, x_1)[x_1, x_2)[x_2, x_3)[x_3, x_4) \ldots \qquad (21)$$

such that:

$$D \equiv (a, x_1)[x_1, x_2)[x_2, x_3)[x_3, x_4) \cdots = (a, b) \qquad (22)$$

So, the division (partition or segmentation) D is defined by a finite, or infinite, sequence of points $\langle x_i \rangle$ within the interval (a, b), having all parts of D, except at most the last part if any, the same length δ, being δ any real number such that $0 < \delta < b - a$. Let x be any point within (a, b) such that $b - x < \delta$. Evidently, the point x can only be a point of the last, or second last, part of D. Therefore, D must have a last part $[x_\varphi, b)$. So, and being the successive parts adjacent, disjoint and of the same length, except at most the last part $[x_\varphi, b)$, D satisfies:

1. D has a first element (a, x_1).

2. D has a last element $[x_\varphi, b)$.

3. Each element of D of the form $[x_i, x_{i+1})$ has an immediate predecessor $[x_{i-1}, x_i)$, or (a, x_1).

4. Each element of D of the form $[x_i, x_{i+1})$ of D has an immediate successor $[x_{i+1}, x_{i+2})$, or (x_φ, b).

Assume now that a part of D of the form $[x_\nu, x_{\nu+1})$ has a finite number ν of predecessors. Being different from $[x_\varphi, b)$, it will have an immediate successor $[x_{\nu+1}, x_{\nu+2})$ in D which will have just one more predecessor than $[x_\nu, x_{\nu+1})$. So, the immediate successor of a part with an immediate successor and a finite number ν of predecessors has also a finite number $\nu + 1$ of predecessors (Peano's

Axiom of the successor [252, p. 1]). Since there is a part of the form $[x_\nu, x_{\nu+1})$ with a finite number of predecessors, for instance the part $[x_1, x_2)$, and each part of D (except the last one $[x_\varphi, b)$) has an immediate successor, we can inductively conclude that all parts of D with an immediate successor, i.e. all parts of D including its second last part, has a finite number of predecessors. Therefore, D has also a finite number of elements (Peano's Axiom of the successor). $\square$

Other more formalized proof based on ordinal numbers can be found in [198]. The impossibility to divide a line of a finite length into an infinite number of parts of the same length is firmly suspected since the XVIII century, at least for some empiricist as G. Berkeley and D. Hume [27, 155, 111].

We will now analyze the possibility of dividing the above interval (a, b) into an infinite number of parts of decreasing length, which is the usual way a finite length is assumed that can be infinitely divided. For this let $S_1 = \langle x_i \rangle$ be an ω-ordered and strictly increasing and convergent sequence of points within (a, b) whose mathematical limit is just b. The sequence of points S_1 defines in (a, b) an ω-Division $D_{1,\omega}$ that can be expressed in the same way as the above case, though now it will be infinite:

$$S_1 = x_1, x_2, x_3, \ldots \tag{23}$$

$$D_{1,\omega} \equiv (a, x_1)[x_1, x_2)[x_2, x_3)[x_3, x_4) \ldots \tag{24}$$

$$(a, x_1) \bigcup_i [x_i, x_{i+1}) = (a, b) \tag{25}$$

$$\lim_i x_i = b \tag{26}$$

An unavoidable feature of ω-divisions is their enormous asymmetry: in our case, any point x arbitrarily close to b will belong to a part $[x_v, x_{v+1})$, so that only a finite number v of parts precedes $[x_v, x_{v+1})$, and an infinite number of them succeed it. In any case, if we remove x_1 from S_1, the remaining sequence S_2 of points will define in (a, b) a new ω-division $D_{2,\omega}$:

$$S_2 = x_2, x_3, x_4, \ldots \tag{27}$$

$$D_{2,\omega} \equiv (a, x_2)[x_2, x_3)[x_3, x_4)[x_4, x_5) \ldots \tag{28}$$

If we remove x_2 from S_2, the remaining sequence S_3 of points will define in (a, b) a new ω-division $D_{3,\omega}$:

$$S_3 = x_3, x_4, x_5, \ldots \tag{29}$$

$$D_{3,\omega} \equiv (a, x_3)[x_3, x_4)[x_4, x_5)[x_5, x_6) \ldots \tag{30}$$

If we remove x_3 from S_3, the remaining sequence S_4 of points will define in (a, b) a new ω-division $D_{4,\omega}$:

$$S_4 = x_4, x_5, x_6, \ldots \tag{31}$$

$$D_{4,\omega} \equiv (a, x_4)[x_4, x_5)[x_5, x_6)[x_6, x_7) \ldots \tag{32}$$

This suggests the following Procedure P:

> Remove from the successive sequences S_1, S_2, $S_3 \ldots$ its first element if the remaining elements define in (a, b) an ω-division.

We will prove now by Modus Tollens that all elements of the initial sequence S_1 can be removed (an inductive proof is also possible [198, pp. 185-186]). Assume that not all elements of S_1 can be removed by the Procedure P. In these conditions, at least one point, say x_v, of S_1 could not be removed. But the sequence:

$$S_{v+1} = x_{v+1}, x_{v+2}, x_{v+3}, \ldots \tag{33}$$

Define in (a, b) an ω-division $D_{v+1,\omega}$:

$$D_{v+1,\omega} \equiv (a, x_{v+1})[x_{v+1}, x_{v+2})[x_{v+2}, x_{v+3})[x_{v+3}, x_{v+4}) \ldots \tag{34}$$

And the same applies to any other point $x_i < x_v$. Therefore, it is impossible that the Procedure P does not remove form S_1 all of its elements, while the remaining ones still define in (a, b) and ω-division.

Obviously, the problem is that if we remove all elements of S_1 we get an empty set of points and it is impossible to divide (a, b) in parts without points to define the division. Note that what has just been demonstrated is not an indeterminacy but an impossibility: the set of points of S_1 that cannot be successively removed without the remaining points defining a ω-division in (a, b) is the empty set; while if we had proved an indeterminacy the set of points of S_1 that cannot be removed without the remaining points defining a ω-division in (a, b) would be an indeterminable non-empty set. The above argument, then, proves the following:

Theorem 4.3 (of the Inconsistent Divisions) *The actual infinite division of any finite real interval is inconsistent.*

We must, therefore, conclude that ω-divisions of finite intervals are inconsistent. And taking into account that ω is the least infinite ordinal, if α is any other infinite ordinal greater than ω, an α-partition of (a, b) will contain an inconsistent ω-subdivision of a subinterval (a, b') of (a, b) [198, Theorem 19]. So, we can end by recalling David Hume's words [155, p. 32]:

> I conclude [...] that no finite extension is capable of containing an infinite number of parts; and consequently that no finite extension is infinitely divisible.

Although Theorem 4.3 applies to both space and time, other independent proof of the inconsistency of the infinite division of space and time intimately related to Zeno Dichotomy are available in [198].

4.7 Discrete or continuous? The problem of change

It takes our brain ≈ 13 milliseconds [260] to process a visual image (α, β, γ and δ movements, and ϕ-phenomenon [89]). Therefore, a continuum of visual images is physiologically impossible. In a movie, motion is also illusory because it consists of a discontinuous sequence of images that we perceive as continuous due to our biological constraints. So, in the same way that a movie is just a discontinuous sequence of images, physical motion could also be a discontinuous sequence of changes in position that our brain and our physical instruments perceive and detect as continuous.

And the same might apply to any other type of change, which is a significant conclusion because change is the most pervasive feature of our ever-changing universe. And at the same time, change is also the most elusive and difficult problem we have ever been faced with [234, 291, 25, 26, 223, 141]. So difficult that it remains unresolved after more than twenty seven centuries trying to solve it (at least in the logical and metaphysical realms).

It may seem exotic to suggest that physics, the science of changes (the science of the 'regular succession of events' in Maxwell's words [218, p. 98]) should be concerned with the problem of change. Physics has been able to identify the magnitudes and establish the mathematical relationships between those magnitudes

in a large number of changes of all kinds: mechanical, thermodynamic, electromagnetic, cosmological, atomic, etc. But change
itself, as such a change, remains unexplained. Neither physics
nor metaphysics have been able to explain how a simple change
in position actually occurs.

Twenty-seven centuries after it was posed by Parmenides, the
problem of change, remains an unsolved problem. With the par-
ticularity that this is a totally forgotten problem. And since forgetting a problem is not the same as solving it, until the problem of
change is solved all physical theories will be provisional. In this
series of papers it will be proved (paper 6) that change is inconsistent within the infinitist framework of the spacetime continuum,
were all solutions have been tried so far. And the reason of that
inconsistency is the lack of immediate successiveness (adjacency)
of the continuum points and instants discussed above. It will also
proved in this series of papers that the problem of change could
find a solution within the scenario of a finite and discrete universe.

Paper 5. A consistent and discrete universe

Abstract. After a short introduction to discrete magnitudes, some significant results on the finiteness of lengths and distances, as well as on the existence of indivisible units of space and time, are proved. An inductive principle is then introduced: the Principle of Directional Evolution, which has the highest empirical evidence and theoretical support. This Principle, which establishes the evolution of the universe in the direction of increasing its entropy, makes it possible to prove the Theorem of the Consistent Universe and nine other subsequent results that clearly point to the discrete nature of space and time. It is shown that space and time, whether real or not, can only be considered as consisting of minimum indivisible units arranged in an adjacent manner. A conclusion incompatible with the spacetime continuum and compatible with certain discrete models such as CALMs (Cellular Automata Like Models), which will be introduced in the article 6 of the series.

Keywords: discrete magnitudes, Principle of Directional Evolution, Theorem of the Consistent Universe, Theorem of Identicality, Theorem of the Formal Dependence, Theorem of the Indivisible Units, Corollary of Discrete Threshold.

5.1 Introduction

Discreteness has been appearing in physics almost unexpectedly. Atomic theory and the ultraviolet catastrophe opened the doors for the first time, and almost simultaneously, to the discrete nature of mass and energy. Clearly, discreteness is not compatible with the infinitist continuum. However, the physics of mass and energy have continued to develop within the framework of the infinitist

63

mathematics of the continuum. In the two previous articles of this series of articles (and in much more detail in [198]) the formal inconsistency of the actual infinity and the formal inconsistency of the infinite division of real intervals (both notions directly involved in the continuum) were demonstrated.

This article recalls the discrete character of some physical magnitudes, as mass or energy. Then it proves that only finite lengths and distances are formally consistent. And, although confirmed by the inconsistency of the actual infinity and the infinite divisibility, it develops several conclusive arguments about the finiteness of lengths, distances and times.

The Principle of Directional Evolution of the Universe is then introduced. This principle establishes that the universe evolves in the direction marked by the increase of its entropy (although what is really important is that there is a directional evolution), for which there is enormous empirical evidence and theoretical confirmation. It is the only principle necessary for the new formal basis of a discrete cosmology. Making use of this principle and the results on discrete sets that were demonstrated in the paper 3 in this series, this 5*th* paper proves nine theorems and corollaries that confirm the discrete nature of space, time and all physical magnitudes.

5.2 Discrete Magnitudes

Until the end of the 19th century, the idea of the infinite divisibility of matter was dominant over the atomistic conception of Leucippus and Democritus. Some relevant authors of the late nineteenth century, such as the mathematician G. Cantor, still denied the existence of atoms [49, p. 78]. But, as is well known, experimental evidence ended up imposing the idea that ordinary matter is discrete, and is made up of indivisible atoms (although atoms can break apart, the fragments are no longer atoms but subatomic particles). Thus, the discrete nature of our observable universe manifested its discreteness for the first time at the level of ordinary matter (matter from now on): every physical object is made up of an integer number of atoms or molecules. An atom, or a molecule, is the smallest possible amount of any given matter. Half an atom of iron, for example, does not make sense. Half an iron atom is a collection of elementary particles, but it is no longer a piece of iron.

According to the Standard Model of Particles, matter is made of a few number of elementary particles: 6 leptons, 6 quarks and 5 bosons, all of them well known (there is an excellent secondary literature and textbooks on this subject, for instance [41, 294, 341, 58, 148, 107, 135, 272, 271, 35, 182, 294, 51, 272, 314, 35, 103, 102, 49]). It is implicitly assumed that all particles of the same type are identical: an electron, for example, from a calcium atom is exactly equal to any other electron from any other atom, or from any cloud of free electrons. This claim will play an important role in later discussions on the discrete nature of the universe. For this and other reasons, this chapter will prove the Theorem 5.5 of Identicality (Section 5.4) that all particles of the same type have the same properties and behave the same way under the same conditions.

The concept of matter, although present in all cultures, is a primitive concept we currently use without a precise definition has been given so far. Notwithstanding, and particularly since the beginning of the XIX century, it has become a fundamental concept of science [12]. Related to the concept of matter is the concept of mass, another primitive concept ("mass is a mess" [244], [166, p. 3]) of which some operational definitions exist [166]. Mass is an attribute of matter of which most of the authors distinguish between inertial mass (a measure of the resistance of matter to acceleration) and gravitational mass (a measure of the gravitational force of matter).

Among all physical magnitudes, mass is the first in the list of physical magnitudes that would have to be of a discrete nature. Indeed, it seems quite clear that:

a) All ordinary matter objects have an integer number of elementary particles.

b) Only a finite number of different elementary particles do exist.

c) The rest mass of all elementary particles of the same type is always the same (Theorem 5.5 of Identicality).

For a given macroscopic body B, let d, n_i and m_i be respectively its number of different atoms, its number of atoms of the type i, and the rest mass of the atoms of the type i. The sum m_{sb} of all rest mass of all elementary component of B will be given by:

$$m_{sb} = \sum_{i=1}^{d} n_i m_i \tag{1}$$

Since a macroscopic body can only have an integer number of

atoms of different types, the above sum of different macroscopic bodies can only vary in a discrete way, and it will always be an integer multiple of certain indivisible units (the rest mass of its corresponding atoms). The electromagnetic assembly of all the atoms of the body B will have a rest mass m_b which will be a certain function f (even the identity function) of the mass m_{sb}.

$$m_b = f(m_{sb}) \qquad (2)$$

And since the variable m_{sb} of the function f can only vary in discrete terms, the rest mass of the different bodies as B can only vary in discrete terms.

The mass-energy relation $E = mc^2$ is almost universally attributed to A. Einstein, but there exist a significant number of historical precedents, from Newton to Poincaré [197]. And, more significantly, there are several non-relativistic ways to derive it [18, 210, 253]. According to the above conclusion on the discreteness of mass, and taking into account that c is a universal constant, the energy E related to the mass m through the speed of light can only vary in discrete terms. So, at least this energy is also a discrete magnitude. Another aspect of energy is the energy involved in the electromagnetic transference of energy. In the first years of the 20th century the works of Gustav Kirchhoff, Ludwig Boltzmann, Otto Lummer, Heinrich Rubens and Wilhelm Wien (among others) led Max Planck [256] to the well known hypothesis that the energy (ϵ) of a electromagnetic radiation is proportional to its frequency:

$$\epsilon = h\nu \qquad (3)$$

and that the energy of an electric oscillator can only take certain values (integer multiples of $h\nu$):

$$h\nu, \ 2h\nu, \ 3h\nu, \ 4h\nu, \ 5h\nu \ \ldots \qquad (4)$$

where $h = 6.62607015 \times 10^{-34} JHz^{-1}$ is Planck constant. According to (4), the energy involved in electromagnetic energy transfers is also a discrete magnitude, without intermediate values between $n\nu$ and $(n+1)\nu$ for any $n \in \mathbb{N}$.

In the case of electric charges there is an elementary charge q whose value is $1.60217662 \times 10^{-19}$ C. This elementary charge is the negative charge of the electron e^- and the positive charge of the proton p^+. On the other hand, the electric charges of quarks are fractions of the elementary charge q:

Quarks down (d), strange (s), and bottom (b): $q/3$.

Quarks up (u), charm (c) and top (t): $2q/3$.

We could also define a new elementary electric charge q_d equal to electric charge of quarks d:

$$q_d = q/3 = 5.340588733 \times 10^{-20} C \tag{5}$$

$$e^- = p^+ = 3q_d \tag{6}$$

$$q_u = 2q_d \tag{7}$$

where q_u represents the electric charge of a quark up. It must be remembered at this point (see paper 4 of this series of articles) that real numbers with an infinite number of decimal places (as $q/3$) are inconsistent when its decimal expansion is considered as a complete totality. In consequence, all indivisible constants and minimal indivisible unities must have a precise and finite number of decimal places (to be precisely determined). In any case, the quantity of electric charges due to the presence of charged elementary particles (electrons and protons) has an indivisible minimum (either q or q_d), and can only vary in discrete terms.

Finally, and again according to quantum mechanics, there is also an indivisible minimum unit for the angular momentum of elementary particles such as electrons or quarks, even though (for other reasons) they are said to have spin $1/2$.

5.3 Finite lengths and distances

As noted in the third article in this series of articles, current infinitism assumes that the list of the natural numbers exists as a complete totality, even though no last number can complete the list. Therefore, it assumes that being incompletable is compatible with being complete. In addition, the successive natural numbers 1, 2, 3,... of that complete list are obtained by adding to an initial unit (the number 1) an infinite number of successive units:

$$1$$

$$1 + 1$$

$$1 + 1 + 1$$

$$1 + 1 + 1 + 1$$

$$1 + 1 + 1 + 1 + 1$$

...

without ever arriving at a number with an infinite number of units, but at an infinite number of numbers each with a finite number of units, each one one unit greater than the previous one. The argument that follows has some relation with that supposed infinite list of the natural numbers, each one unity greater than the previous one, but all of them finite. It proves that in the Euclidean space $\mathbb{R}^3$ every line (whether or not straight) with two endpoints, either open or closed at those endpoints, has a finite length. An immediate consequence is that even in a space of infinite extension it is impossible to find two points separated by an infinite distance. Let us then demonstrate the following important theorem:

Theorem 5.1 (of the Finite Lengths) *In the Euclidean space* $\mathbb{R}^3$ *every line with two endpoints has a finite length.*

(Although the first part of the following proof is the same as the proof of Theorem 4.3, it is repeated here with some variation.)

Proof: Let AB be any line in the Euclidean space $\mathbb{R}^3$, and $\lambda > 0$ any finite length. Let $\mathbf{P} = AP_1, P_1P_2, P_2P_3 \ldots$ be a partition of AB all of whose parts have the same finite length $\lambda > 0$, except at most the last one, if any, whose length can be less than λ. A point X such that $XB < \lambda$ can only be in the last, or second last, part of $\mathbf{P}$. So, $\mathbf{P}$ has a last part $P_\phi B$. For every P_i in $\mathbf{P}$ it holds: any point Y of the segment AP_i such that $YP_i < \lambda$ a can only belong to $P_{i-1}P_i$, or to AP_1. And any point Z of the segment P_iB such that $P_iZ < \lambda$ can only belong to P_iP_{i+1} or to $P_\phi B$. In consequence, each part of $\mathbf{P}$ of the form $P_{i-1}P_i$ has an immediate predecessor $P_{i-2}P_{i-1}$ (or AP_1) and an immediate successor P_iP_{i+1} (or $P\phi B$). In additions, AP_1 has an immediate successor P_1P_2, and $P_\phi B$ has an immediate predecessor $P_{\phi-1}P_\phi$. So, $\mathbf{P}$ has a first element AP_1, a last element $P_\phi B$, and each element has an immediate predecessor (except AP_1), and an immediate successor (except $P_\phi B$). Let us suppose there exists a part in $\mathbf{P}$ with an immediate successor and a finite number n of predecessors. The immediate successor of this part will also have a finite number $n + 1$ of predecessors (Peano's Axiom of the successor [252, p. 1]). Since P_1P_2 has an immediate successor P_2P_3 and a finite number of predecessors, just one predecessor AP_1, we can inductively conclude that all parts of $\mathbf{P}$ with an immediate successor, i.e. all parts of $\mathbf{P}$ until its second last part have a finite number of predecessors. So, $\mathbf{P}$ has a finite number of elements (Peano's Axiom of the successor). And being finite the sum of any finite number of finite lengths, AB has finite length. □

Several immediate corollaries of Theorem 5.1 can now be easily proved [198, pp. 178-180]. Some of then are the following ones:

Corollary 5.1 (of the Closed Lines) *In the Euclidean space $\mathbb{R}^3$, any closed line has a finite length.*

Corollary 5.2 (of the Finite Distances) *In the Euclidean space $\mathbb{R}^3$ the distance between any two of its points is always finite.*

Corollary 5.3 (of Finite Connections) *In the Euclidean space $\mathbb{R}^3$ it is impossible to join any two of its points by a line of infinite length.*

Corollary 5.4 (of Infinite Lengths) *In the Euclidean space $\mathbb{R}^3$ lines of infinite length are inconsistent.*

Where line stands for any type of line, be it, or not, a straight line. An immediate consequence of the above results is that if we remove from the supposed infinite $\mathbb{R}^3$ all couples of points separated by a finite distance, the result can only be either just one point or the empty set. Under these conditions it is not at all clear what meaning an infinite universe could have. In any case, according to the above proved results, and taking into account the spacetime continuum is modeled by the set $\mathbb{R}^4$, we can enunciate the following:

Theorem 5.2 (of Finite Distances and Durations) *In the space-time continuum, the distance between any two points and the time elapsed between any two instant is always finite.*

This agree with Huby's opinion [152, p.121]:

> That the universe, if real, must be finite in both space
> and time.

In addition to being finite, the intervals of space and time cannot be divided into an actual infinite number of parts (as was proved in the paper 4 of this series of papers). It is only possible to do it in a finite number of parts. And we have to consider the following two alternatives, which are exhaustive and exclusive:

1. There is an indivisible unit defining length intervals and time intervals.

2. There is not indivisible units for at least one of the magnitudes length and time.

In principle, and in an abstract world, both alternatives would be possible. Indeed, it is always possible to consider a number less than any given number, in the same way that it is always possible to consider a natural number greater than any given natural number (potential infinity). But in the observable universe, and taking into account all the theoretical and empirical knowledge accumulated (especially since the Scientific Revolution [172, 209, 215, 296, 316]), things seem to be different, precisely because they are observable and because they have always been observed subject to the same basic regularities (physical laws). As will be seen in the next section, the very existence of physical laws follows from the Principle of Directional Evolution, and from those laws it is possible to prove the existence of indivisible minima of space and time.

5.4 The universe is consistent

The observable universe contains billions and billions of objects of the same type: galaxies, stars, planets, minerals, chemical elements... uniformly distributed and of very different ages. This is only possible if the same things have always been happening in the universe, and in such a way that under the same conditions the same consequences have always resulted. Something that has been suspected for at least a couple of centuries: the naturalists of the 19th century already assumed the Principle of Actualism Uniformism which suggests the same conclusion:

The laws of nature are the same in all places and times.

Even this general principle can be deduced in formal terms from another even more general principle. For, in effect, the observable universe has been producing the same type of objects throughout its history (over 13.8 billion years): there are planets, stars, galaxies etc. of all ages. And all these objects have been evolving in the same way. In addition, there is a law of thermodynamics (though with numerous versions), the Second Law of Thermodynamics, that also points in the same direction regarding the evolution of the heat-energy interconnections. All this, and taking into account that for most of its history there were no rational observers in the universe, amply justifies proposing the following:

Principle 5.1 (of the Directional Evolution) *The universe al-ways evolves independently of its rational observers and in the same direction of increasing its global entropy.*

Where entropy can be replaced with isotropy [195]. This primor-dial directionality has made possible other local directional evolu-tions that seem to go in the opposite direction: the generation and evolution of open systems that exchange matter and energy with their environment. In these systems there is a notable decrease in entropy, but in return there is an even more notable increase in the entropy of their surroundings, so that the final balance is a directional evolution of the universe in the sense the above Principle of Directional Evolution establishes. This is the case of crystalline mineral and, about all, the self-organizing systems we call living beings [188, 189].

Along with the Principle of Directional Evolution, we will also assume the following definitions and fundamental principle:

Definition 5.1 (of the Consistent Sets of Laws) *A set of physi-cal laws is consistent if under the same conditions it always leads to the same results.*

Definition 5.2 (of Discrete Magnitudes) *A magnitude is discrete if any of its values is an integer multiple of a minimum, indivisible and invariant value.*

Principle 5.2 (of Discrete Magnitudes) *Some physical magnitu-des are discrete.*

There is also enormous empirical evidence for the existence of these discrete magnitudes, as in the case of electromagnetic en-ergy, mass, electric charge, angular momentum of electrons, of atomic nuclei, or of atoms. As will be seen below, from both princi-ples we immediately deduce some fundamental results that detail the way in which the universe evolves:

Theorem 5.3 (of the Consistent Universe) *The universe evolves independently of its observers and under the control of a unique set of invariant and consistent physical laws.*

Proof: If the physical laws governing the evolution of the universe were not an invariable set of consistent laws, changes would occur with the same frequency in all directions, and no progress would be possible in any of them. So, directional evolution would not be possible, which goes against the Principle 5.1 of Directional

Evolution. So then, the universe evolves under the control of a unique set of invariant and consistent physical laws. □

In reality, and as incredible as it may seem, the Theorem of the Consistent Universe is absolutely necessary. To understand this necessity, it is enough to remember a very famous phrase of a very famous scientist, A. Einstein, which is found in many physics books [88, p. 315]:

> One can say: The eternal incomprehensible of the world is its comprehensibility.
>
> Popularized as: The fact that the universe is understandable is a miracle.

But, for the reasons given above, a universe that produces scientists (through a process that here on Earth has lasted several billion years) can only be a universe that evolves in a directional way and, consequently, under the control of a single set of invariant and formally consistent laws. That is, a universe understandable in terms of constant and consistent regularities. The miracle would be, then, that a universe with scientists were not understandable.

Another thing is that the mathematical language used to explain the universe is appropriate. In this series of articles we are showing that it is not. That disagreement would explain why it is so difficult to understand what should not be so difficult to understand.

Theorem 5.4 (of the Discrete Magnitudes) *If the mathematical output of a physical law is a discrete magnitude, its definition cannot contain continuous variables.*

Proof: If the mathematical definition of a physical law contains any continuous variable, its output (result) will always be continuous, not discrete as required for a discrete magnitude. □

Theorem 5.5 (of Identicality) *All particles of the same type have the same properties and behave the same way under the same conditions.*

Proof: It is an immediate consequence of Definition 5.1, Theorem 5.3 and Corollary 5.5. □

Theorem 5.6 (of Formal Dependence) *No concept defines itself; no statement proves itself; no physical object is the cause of itself; and no cause is the cause of itself.*

Proof: If propositions could prove themselves, then any proposition could be proved, and consistent sets of laws would be impossible, which goes against Theorem 5.3. If concepts defined themselves, their meanings would be inaccessible to human knowledge, and they could not be used to establish the natural laws that we can only establish with those concepts, which also goes against Theorem 5.3. If physical objects, processes or causes were the cause of themselves, then anything could exist and anything could happen, so that the directional evolution of the universe would be impossible, which goes against the Principle of Directional Evolution. □

It is then clear that in our observable universe, under the same conditions the same consequences result. Another thing is that those results may seem strange or paradoxical to us. In that case, the corresponding laws will always produce the same strangeness or perplexity, as occurs with some aspects of quantum mechanics. Moreover, if the universe worked in a way similar to cellular automata, perhaps two types of laws could exist: the basic laws of the automaton (established before starting the automaton) and the laws that emerge from its evolution and govern the relationships between the objects, also emerging, from the automaton evolution (see next paper 6)

From now on, and for the sake of simplicity, the directional evolution of the observable universe will be referred to simply as the evolution of the universe. No exception to the consistent evolution of the universe, on the other hand, is known. It is the reason why the universe allows itself to be described with formal and computational languages. Whether or not the current infinitist mathematical language of physics is the most appropriate is another matter. And surely it is not because it is based on an inconsistent axiom (the Axiom of Infinity).

5.5 Indivisible units of space and time

The speed c of light in the vacuum[1] is one of the universal constants of physics of which we have the greatest empirical evidence. It is the speed of an object (as a photon) that takes a Planck time

---

[1]The vacuum is used here as the carrier medium for the physical fields

to traverse a Planck length:

$$c = \frac{l_p}{t_p} = \frac{1.616255 \times 10^{-35}\,m}{5.391247 \times 10^{-44}\,s} = 299792423\,m/s \tag{8}$$

The speed c of light in a vacuum can also be defined in terms of other pair of universal constants, the electric permittivity ϵ_o and the magnetic permeability μ_o of the vacuum:

$$c = \frac{1}{\sqrt{\epsilon_o\,\mu_o}} = 299792423\,m/s \tag{9}$$

(In SI, c is defined as 299792458 m/s because a meter is defined in the SI as the distance light travels in $1/299792458$ s). According to (8) and (9), it is clear that the speed c of light in the vacuum is a universal constant. And it is not only the speed of electromagnetic waves through the vacuum, it is also the speed of the propagation through the same vacuum of other perturbations of other physical fields, for example the propagation of gravitational waves. Thus, l_p and t_p could be respectively the indivisible and invariant minimum of space and time

On the other hand, from (8) and (9) it immediately follows:

$$t_p = l_p/c = l_p\sqrt{\epsilon_o\,\mu_o} \tag{10}$$

which is a rather enigmatic relation between the possible unit of discrete time (qutit) and the possible unit of discrete space (qusit) defined through two universal constants, in principle unrelated to metrical properties of space and time, namely the electric permittivity and the magnetic permeability of the vacuum (free space?)

The electromagnetic spectrum is considered in contemporary physics as continuous and (virtually) infinite, for instance [332, p. 891]:

> The wavelengths of electromagnetic waves have no inherent upper or lower bound.

But since the wavelength of any wave is defined as the distance between two successive points in the same state of vibration, then, and according to the Theorem 5.1 of the Finite Lengths, all wavelengths will be finite. And if the set of all possible wavelengths exist, then, and according to Theorems 3.10 and 3.11, that set can be discretely ordered, with a minimum and a maximum.

I will now demonstrate some important results from the perspective of a finite and discrete universe:

Theorem 5.7 (of Discrete Space and Time) *Space and time are discrete, each with an indivisible and invariant minimum.*

Proof: Since space and time are involved in the definition of discrete physical magnitudes, according to Theorem 5.4 they must be discrete entities, each with an indivisible and invariant minimum, otherwise the defined discrete magnitudes could not be consistently discrete. □

From now on, the indivisible and invariant minimum units of space and time will be referred to simply as minimum units. The existence of these minimal units of space and time, deduced in purely formal terms, supports the idea that Planck length and Planck time, although deduced from dimensional equations, could also be indicating the existence of minimum unit of space and an a minimum unit of time. These minima impose certain restrictions to experimental and theoretical physics. One could not measure, for example, the speed of light over a distance $d < l_p$, nor the duration t of events such that $t < t_p$. These limitations have been theoretically confirmed and experimentally tested (see, for instance, [10, 146, 254, 236]). And here, it is formally proved in the form of the following:

Theorem 5.8 (of the Discrete Threshold) *The laws of physics do not apply in spaces smaller than the minimum unit of space nor in times smaller than the minimum unit of time, both being of non-zero extension (duration).*

Proof: If the laws of physics could be applied in spaces smaller than the minimum unit of space, then the space involved in the mathematical functions whose outputs have to be the discrete values of the magnitudes defined by those functions, would not be an integer multiple of that minimum unit of space, it could be any value, in which case the outputs of those discrete functions would not have the discrete values they would have to have: only integer multiples of the corresponding minimum unit of the magnitude defined by each of those discrete functions. The same argument holds for time. □

As noted above, although the Theorem 5.8 of the Discrete Threshold has not been explicitly stated in contemporary physics, its statement has broad theoretical and empirical support. As will be seen throughout this series of articles, it is a fundamental result for the construction of discrete models of the universe.

Corollary 5.5 (of the Physical Laws) *The laws of physics apply to all regions of space and time, provided they are not less than their respective minimum units.*

Proof: It is an immediate consequence of Theorems 5.3 and 5.8. □

Theorem 5.9 (of non-Extensive Points) *The points (instants) of the spacetime continuum have not extension (duration).*

Proof: Suppose that the points of the spacetime continuum have an extension of δ meters, being δ a real number greater than zero. Let AB and CD be the lengths of any two lineal intervals of that continuum such that:
$$AB < CD \tag{11}$$
Since the number of points of AB and CD is the same, just $2^{\aleph_0}$, we would have:

$$AB = 2^{\aleph_0} \times \delta \text{ m} = 2^{\aleph_0} \text{ m} \tag{12}$$
$$CD = 2^{\aleph_0} \times \delta \text{ m} = 2^{\aleph_0} \text{ m} \tag{13}$$
$$\therefore AB = CD \tag{14}$$

which contradicts (11). Therefore, points cannot have an extension greater than zero. For the same reasons, instants cannot have a duration greater than zero. □

Theorem 5.10 (of Application of Physical Laws) *In a consistent universe, the laws of physics cannot be applied to a point, nor during an instant of the spacetime continuum. Neither can they be applied to an interval of points nor during an interval of instants of the spacetime continuum.*

Proof: Since, according to Theorem 5.9, points (instants) have no extension (duration), to apply a physical law to a point during an instant is to apply that law to no space during no time. Nor can it be applied to an interval of space during an interval of time of the spacetime continuum because if the interval has an infinite number of points (instants) then it is inconsistent (Corollary 3.4); and if it has a finite number of points (instants) then it has no extension (duration) (Theorem 5.9), and it would be the same as applying it to a point during an instant, i.e. applying it to no space during no time. □

Theorem 5.11 (of Space and Time Intervals) *All space and time intervals in which apply the laws of physics are integer multiples of their respective minimum units.*

Proof: It is an immediate consequence of Definition 5.2 and Theorem 5.7. □

Theorem 5.12 (of Adjacency) *No space exists between any two successive space minimum units, and no time elapses between two successive time minimum units.*

Proof: Let AB and CD be two successive space minimum units (simplified to a one dimensional version) and assume they are not adjacent, i.e assume that $0 < BC$. BC must be less than the space minimum unit, otherwise AB and CD would not be two successive space minimum units. Consequently, AD would not be an integer multiple of the space minimum units, which is impossible according to Definition 5.2 and Theorem 5.7. □

Corollary 5.6 (of Finite Space and Time) *Every space (or time) interval is finite and can only be divided into an integer number of adjacent space (or time) minimum units.*

Proof: It is an immediate consequence of Theorems 4.3, 5.1 and 5.12. □

Note that in a discrete reality, in which both space and time were discrete, no object accessible to study (e.g. elementary particles) could be smaller than a minimum unit of space (qusit), nor last less than a minimum unit of time (qutit). The existence of these limits for the intervals of space (Planck length) and time (Planck time) are also assumed in quantum mechanics, and their existence can be proved in semi-formal terms from Heisenberg's Principle of Uncertainty [197, pp. 269-272]. There would exist, besides, a maximum velocity of one qusit per qutit, so the Second Principle of special relativity would not be necessary, but an immediate consequence of the discrete nature of space and time.

As indicated in articles 1 and 4 of this series, the primary and secondary physics literature is replete with expressions that reveal the scant interest of physicists in the formalism of their infinitist mathematical language: *adjacent points*, *contiguous points*, *point to point*, and so on. But in the spacetime continuum there are no adjacent points, nor contiguous points, nor is it possible to go point to point. Between any two points of this spacetime continuum there always exists the same non-countable infinite number of different points, just $2^{\aleph_0}$, as many as in the whole three-dimensional universe.

An immediate consequence of the above theorems and corollaries is that physical objects, if they have size (could a physical object not have size?), cannot be point-like, nor have the same structure as a point because points neither have size nor, consequently, have structure. Nor can something physical be concentrated in a point without making it disappear, because points do not occupy space: something concentrated in a point would not occupy any space. However, authors who need no introduction have written texts such as:

> ... a beam of light emanating from a point... consists of a finite number of energy quanta, localized at points of space... (A. Einstein, quoted in [41, pp. 45-46].

> ... the interpretation of the photon as a point like structure... (A. Einstein, quoted in [41, p. 46].

> ... in the position to concentrate energy upon a single point in space... (M. Planck, quoted in [41, p. 46].

I think that bothering a bit with such reflections would have quickly led to considering the discrete alternative for space and time.

Paper 6. Discrete models

Abstract. This paper 6 of the series begins by recalling the old problem of change, one of the most basic problems of physics, but one that physics has not dealt with for centuries. First, the problem of change is posed in formal terms, and then it is proved that it has no solution in the spacetime continuum. It is formally demonstrated that, in effect, change is an impossible process in this continuum, the reason for this impossibility being the lack of immediate successiveness (adjacency, contiguity) between the elements of the continuum (points and instants). A discrete model inspired by cellular automata (CALM) is then proposed where the problem of change could finally be solved. It will be the same model that in subsequent papers of this series will be proposed as an initial step in the search for a discrete and finitist model for the observable universe.

Keywords: problem of change, canonical change, spacetime continuum, CALM, cellular automata, qusit, qutit.

6.1 Introduction

The protagonist of this Part 6 is the old problem of change, which was raised more than twenty-seven centuries ago [234, 291, 25, 26, 223, 141]. A problem that, although forgotten in modern science, has not yet been solved. An oversight that is particularly important in physics, the science of change (the science of the regular succession of events in J. C. Maxwell words [218, p.98]); the science that should be the most interested in its solution. Indeed, it is not a satisfactory fact for a science as physics to have to admit its inability to explain the problem of change; its inability to explain how a simple change of position of an object in uniform

79

motion occurs.

As is well known, the problem of change is related to Zeno's Dichotomy [190], a dichotomy to which several solutions have been proposed from different areas of mathematics and physics [125, 126, 342, 127, 129, 128, 222, 221, 246, 6, 266, 290, 154, 300], but the problem of change itself remains unsolved. Forgetting a problem is not a way to solve it. And not being able to explain how a simple change of position of an object in uniform motion occurs should be pointing to a basic deficiency in the model used to explain the physical world. That model has classically been based on a time and on a space essentially continuous (see paper 4), which basically coincides with the modern spacetime continuum, the supposedly infinitist structure of space and time in which all solutions to the problem of change have been sought.

The simplicity of the problem and the sterile search for solutions for more than twenty-seven centuries have surely influenced the abandonment of the search and the pretense that the problem does not exist. But the problem exists, and it is very basic. The behavior of science with the problem of change is really shameful. As the only justification, it could be said that our sensory perception of the physical world deceives us: it shows us a continuity that could be discontinuous (think of a movie and its frames). But, in any case, this discontinuity should have been explored. Especially since the invention of cinema. As will be seen in this paper, the problem of change could be solved in a model where space and time are discrete instead of continuous, which, naturally, would be confirming the conclusions obtained in the precedent articles of this series of articles.

As just indicated, physics models the physical world as a continuous world probably because we sensorially perceive it as a continuous world. The problem is that this perceived continuity is actually illusory because of the way our brain constructs the images we see: it lasts a time greater than zero (≈ 13 ms [260]) to process each visual image (α, β, γ and δ movements, and ϕ-phenomenon [89]), so that a continuum of visual images is physiologically impossible, they will always be separated by a time interval greater than zero. In the same way that a movie is a discontinuous sequence of images perceived as a continuum, natural motion could also be a discontinuous sequence of changes of position that, for the same reason as in a film, is perceived as continuous by our brain. This illusory continuum (the impossibility of a sensory perception of nature discreteness) is surely behind our attempts to explain the physical world in term of the spacetime continuum.

The discrete nature of space and time would surely open the door to a discrete interpretation of special relativity (the science of the spacetime continuum) in terms of apparent, not real, space contractions, time dilations and local simultaneity (lack of universal simultaneity) [197].

The problem of change has been forgotten by physics, surely because of its unconditional ascription to the spacetime continuum. Indeed, the development of contemporary theoretical physics has occurred exclusively within the framework of infinitist mathematics, and with a total lack of interest in the formal consistency of the Hypothesis of the Actual Infinity that underpins infinitist mathematics. Fortunately, experimental physics can only be discrete and finitist, and requires theoretical models to be adapted to its results. Dealing with the problem of change would have had a double benefit: check the inconsistency of its infinitist mathematical language, and solve the problem of change itself. To have faced the problem of change would surely have had the consequence of discovering that the most appropriate language for physics is not infinitist mathematics but finitist computational language, or a new type of undeveloped discrete mathematics. But it's never too late.

After posing the problem of change in formal terms, this paper demonstrates that it cannot be solved within the framework of the spacetime continuum, the only framework in which its solution has so far been sought. And it cannot be solved precisely because of the lack of immediate successiveness (lack of adjacency between points and between instants) in the spacetime continuum. It is then demonstrated that the problem of change could be solved within the framework of a discrete space and a discrete time, which here will be a model similar to cellular automata (CALM, cellular automata like model). The discrete solution of the old pre-Socratic problem would confirm the need for a discreet model to explain the physical world, because the physical world is essentially a consistent and constantly changing world.

(The text of the following two sections is an up-to-date summary of [198, pp. 329-338] and [197, pp. 571-583].)

6.2 The problem of change

If Ob is a physical object, we will say Ob changes causally from the state S_a to the state S_b if there exist a set of (physical) laws L such that, under the same conditions C, and as a consequence of those laws and conditions, the state of Ob is S_a at the instant t_a, and S_b at an ulterior instant t_b, symbolically:

$$S_a \mapsto S_b : \quad L(S_a, C, t_a) = (S_b, t_b) \tag{1}$$

Here we will only deal with causal changes defined according to (1). They will be referred to simply as changes.

The change $S_a \mapsto S_b$ can be direct, without intermediate states. In such a case, it will be referred to as *canonical* change. It can also be the result of an ordered sequence of canonical changes:

$$\langle S_a \mapsto S_b \rangle : \quad S_a \mapsto S_1 \mapsto S_2 \mapsto \cdots \mapsto S_v \mapsto S_b \tag{2}$$

Note that, except S_1, each element S_n of $\{S_i\}$ must have an immediate predecessor S_{n-1} (symbolically $S_{n-1} < S_n$) so that S_n can be *causally* derived from S_{n-1}:

$$\forall S_{1 < n \leq b} : \quad L(S_{n-1}, C_{n-1}, t_{n-1}) = (S_n, t_n) \tag{3}$$

The objective of the discussion that follows is the analysis of the canonical changes, whether or not they are part of a sequence of canonical changes. We will begin by proving the following two theorems:

Theorem 6.1 (of the Canonical Changes) *Every change is either a canonical change of a discrete and finite sequence of canonical changes.*

Proof: Let $S_a \mapsto S_b$ any change. If it is not a canonical change it will be a sequence of changes. A sequence that cannot be densely ordered (Theorem 3.7). Therefore, it will be a sequence with a first change; a last change; and each change (except the first) will be immediately preceded by another change and will be immediately followed by another change (except the last). It will therefore be a discrete sequence of canonical changes, which can only be finite (Theorem 3.9). □

Theorem 6.2 (of Change) *Canonical changes are instantaneous and then impossible in the spacetime continuum.*

Proof: Let $S_a \mapsto S_b$ any change of any object Ob and suppose it lasts for any time $t > 0$. Let t' be any instant in the interval $(0, t)$. If at t' the state of Ob is S_a, the change has not yet begun, and its duration will be less than t. If the state of Ob at t' is S_b, the change will have already ended and its duration will also be less than t. Therefore the duration of the change will be less than any real number t greater than zero. The duration of the change cannot be negative either because in that case S_b would be prior to S_a. Therefore the duration of the change has to be non-negative and less than any real number greater than zero. That is, it must be zero. The change must be instantaneous, which in spacetime continuum is only possible if both states coexist in the same instant, because in the spacetime continuum between any two different instants always elapses a time greater than zero. Now, if both initial and final states coexist in the same instant it is not possible to establish which state is the cause of the other. Therefore, canonical change is impossible in the spacetime continuum. □

In the last section of this article we will analyze whether canonical changes are possible in a discrete space-time model. And before that, in the next section, some interesting questions about continuous versus discrete reality will be raised.

6.3 Discrete versus continuous

The main objective of this series of articles, to propose a discrete reality as opposed to the assumed continuous reality, may seem exotic and unnecessary. Although a discrete reality would be much simpler than the continuous reality founded on the spacetime continuum that in our days, and since the beginning of the history of science, is naturally assumed. As indicated above (and especially in the article 4 of the series), something must have influenced the fact that we perceive the world as essentially continuous, although we know that this apparently continuous perception is in fact a deception of ours brains (similar to that of cinematography).

Leaving aside the fact that the universe is consistent (Theorem 5.3 of the Consistent Universe) and the spacetime continuum inconsistent (Theorem 3.8 of the Inconsistent Continuum), we will have to admit that it is not the same to explain the evolution of $\approx 2.66 \times 10^{185}$ qusits[1] as that of $2^{\aleph_o}$ points, with the additional difficulty that any area of the universe, however small, has the same number of points as the entire universe. In addition, some interesting questions that physics has ignored throughout its history may also be raised:

---

[1]The total number of qusits in the observable universe if they where of a Planck volume.

1. If space can deform, expand and vibrate, is it not physical? (see paper 11). In these conditions, how does physical space relate to the geometric points of the continuum?

2. If space is physical, is it transparent to physical objects?

3. If space is physical, what is the physical reality of the points? And how does ordinary matter relate to those points?

4. How is it possible that, for example, in a linear space of one millionth of a millimeter and during one millionth of a second as many virtual particles are created as in the entire three-dimensional universe during its entire history of more than 13.7 billion years?

5. How is the irreversible and directional geological record possible in a reversible and non-directional spacetime continuum?

And above all:

6. How can a consistent universe be incessantly changing if change is inconsistent in the spacetime continuum?

7. How is it possible that physics pretends to explain a physical world in continuous change without previously solving the problem of change?

8. Will these questions deserve the attention of the hegemonic infinitist streams of thought in modern physics?

As will be seen in the next section, most of these questions, and many others not properly addressed by modern physics, could find a simple answer in discrete and finitist models of space and time.

6.4 A discrete model: cellular automata

Cellular automata like models (CALMs) provide a new interesting perspective to analyze the way the universe could be evolving. In particular it provides a discrete spacetime model in which a new analysis of the incomprehensible oddities of contemporary physics, including change, would be possible. As we will see in the next short discussion, twenty seven centuries after it was posed, the old problem of change could find a first consistent solution in the discrete spacetime of CALMs.

In CALMs, space is made up exclusively of minimal indivisible units: cells (qusits). Time is also made up of a sequence of successive indivisible minimum units: qutits. There is no extension between a qusit and its immediate successor in any spatial direction. Similarly, no time elapses between a qutit and its immediate successor. Each qusit can exhibit different states, each defined by a certain set of variables. The states of all qusits change in unison, simultaneously, in each successive qutit according to the laws that drive the evolution of the automaton. Once changed, the state of each qusit remains unchanged for one qutit. In what follows we will assume that this is the case, although instead of one qutit, the state of each qusit could also remain unchanged for a certain integer number $n \geq 1$ of qutits. Note that the problem of change has not yet been solved: it will be necessary to explain how the successive changes of state of each qusit occur. A consistent way of explaining how such changes could occur is proposed on page 129.

Let u, v, c, ... z be the set of variables that define the state of each qusit of a certain CALM A. Let us represent the nth state of each qusit x_i by $x_i(u_{i,n}, v_{i,n}, ... z_{i,n})$, where $u_{i,n}$, $v_{i,n}$ ... $z_{i,n}$ denote the particular values of the state variables of x_i at the nth qutit. Let finally L be the set of laws driving the evolution of the automaton, including the laws that relate the different state variables to each other. L determines the way each qusit x_i changes from a qutit to the next one, taking into account the state of x_i as well as the state of any other qusit with which it interacts, which may include all qusits. All these current states define the conditions C_i under which the laws L determine the state of each qusit in the next qutit, that is, the laws that determine the change that each qusit undergoes in each successive qutit.

The automaton engine changes the state of every qusit at each qutit and maintains it just for one qutit. Thus we can write for each particular qusit x_i:

$$L(x_i(u_{i,n} \ldots, z_{i,n}), C_n, t_n) = (x_i(u_{i,n+1} \ldots, z_{i,n+1}), t_{n+1})$$

$$L(x_i(u_{i,n+1} \ldots, z_{i,n+1}), C_{n+1}, t_{n+1}) = (x_i(u_{i,n+2} \ldots, z_{i,n+2}), t_{n+2})$$

$$L(x_i(u_{i,n+2} \ldots, z_{i,n+2}), C_{n+2}, t_{n+2}) = (x_i(i, u_{n+3} \ldots, z_{i,n+3}), t_{n+3})$$

$$L(x_i(u_{i,n+3} \ldots, z_{i,n+3}), C_{n+3}, t_{n+3}) = (x_i(u_{i,n+4} \ldots, z_{i,n+4}), t_{n+4})$$

$\ldots$

Certain sets of qusits could remain grouped with the same configuration through the successive qutits. They could be said CALM's objects. It is significant that the operation of a CALM is similar to that of a computer: its internal clock defines the indivisible units of time in which all operations and updates occur. And remember that computers are man-made machines capable of simulating physical phenomena.

Being both space and time discrete, each qutit t_n has an immediate predecessor t_{n-1} and an immediate successor t_{n+1}, so that no other qutit elapses neither between t_{n-1} and t_n nor between t_n and t_{n+1}. Or in other words: no time passes between any two successive qutits. This simple characteristic of CALMs (together with the permanence and interaction modes explained later in this document) suffices to solve the logic problem of change because discrete spacetime allows instantaneous changes: the state A_n at qutit t_n changes to A_{n+1} at the next qutit t_{n+1}, being zero the time elapsed between t_n and t_{n+1}. It could be said that all qusits of a CALM are updated simultaneously at each qutit. In the case of the points and instants of the spacetime continuum, things are different because between any two of its points (instants), whatever they are, there are other $2^{\aleph_o}$ different points (instants), so that none of them has an immediate successor, which makes it impossible for change to occur.

On the other hand, we must not forget that our sensory perception of the physical world is continuous. And this is why we are used to think in terms of a spacetime continuum. So far, our only way of thinking. All our models of the physical world have assumed the physical world is a continuous world. It is then almost inevitable to extrapolate this way of thinking to any new discrete paradigm, which obviously would be catastrophic. To think in (physical) discrete terms will surely require a long process of reeducation.

An electron, for instance, could be in the state S_1 at a certain instant t_1, and in another state S_2 at another posterior instant t_2, without ever being in any intermediate state between S_1 and S_2 (quantum jump). It is therefore a canonical change. In the spacetime continuum the interval (t_1, t_2) must always be greater than zero and during that time the electron can be neither in the state S_1, nor in the state S_2 nor in any other imaginable state. Therefore, it must cease to exist for a time greater than zero. It must disappear at t_1 and reappear at t_2. In the discrete space and time of a CALM all we have to do is to consider two successive qutits, t_1 and t_2. At t_1 our electron would be in the state S_1, and at t_2 in the state S_2, so that no other qutit passes between t_1 and t_2. But we must recognize this is an incomplete explanation, as we will immediately see.

Though instantaneous changes are possible in such discrete space-times, it is very difficult to grasp the idea of instantaneous changes. Indeed, how can a change be instantaneous? If the change results from a process (the process of change) and that process has a zero duration, the process has no existence and the change remains impossible. We arrive at the starting point of Zeno's paradoxes, immediate consequences of the impossibility of change. But changes exist, they do not stop happening. Therefore, everything indicates that we need a new paradigm about the intimate constitution and functioning of the physical world at its most essential scale, even beyond the atomic scale.

The directional evolution of the universe demonstrates that this evolution is subject to a consistent set of rules, physical laws, (Theorem 5.3 of the Consistent Universe). So, canonical changes have to be consistent processes, and then instantaneous. The problem is that we have no idea how that is possible. As a very adventurous hypothesis, it could be proposed that qusits have two modes of existence:

1. Permanence mode: the state of each qusit remains unchanged at least for one qutit. This would be the only perceptible state of qusits.

2. Interactive mode: all qusits update synchronically their respective states through appropriate processes lasting at least one qutit.

Although, in accordance with what was said above, the problem of change will now appear in the terms of these changes of modes. So, we would have to admit that the interactive mode is simultaneous with the permanence mode, although it remains in an imperceptible background (such as computer applications running in the background) that change to the perceptible mode at each successive qutit (or something similar). Or in other words, the perceptible state of the qusits would coexist with the interactive mode of updating (changing) that will define the next perceptible state, which will become perceptible in the next qutit. The newly substituted perceptible states being the new source of interactions of the new interactive mode.

It is interesting to remember that in computers very rapid changes occur in the contents of their discrete units of memory (their qusits) at the rate set by the successive discrete units of time of their internal clocks (their qutits). The contents of these discrete units of memory are updated by successive qutits, and with them the various devices controlled by each computer.

One could argue that the same could happen in the spacetime continuum: the content of each point (or group of points) is updated at each instant (or group of instants). The problem here is that no point (or group of points) has adjacent points (or adjacent groups of points): there are neither adjacent points nor groups of adjacent points (no group has a last point adjacent to the first point of the next group). And the same is true for instants (or groups of instants). So the above possible discrete solution of the problem of change cannot be applied to the spacetime continuum.

What has just been presented is not the solution of the problem of change, but a way of solution based on the discreteness of space and time. The only clear thing is that in the spacetime continuum this solution is not possible. Nor is any other because in the spacetime continuum canonical changes are impossible. (Theorem 6.2).

From the point of view of CALM model, it would be interesting to analyze its compatibility, or even its formal relationships, with the implicit order proposed by David Bohm [38]. In any case, and by way of example, assume that:

- The universe has 7.6564×10^{196} qusits.
- The universe contains 10^{80} elementary particles.
- Each particle is defined by p variables
- Each particle is, somehow, present in each qusit.

Let U be a tridimensional CALM of 7.6564×10^{196} qusits in which the state of each qusit is defined by $p \times 10^{80}$ state variables. If it were possible to simulate U, perhaps we would observe the self-organizing and evolution of an object similar to our universe.

The CALM U would be incomparable less complex than, for instance, any matrix of infinite elements (which are usual in mathematics and theoretical physics). We could model the universe, provided we know the basic laws that make it evolve. In this circumstances, to simulate does not means to reproduce the exact history of the universe: recursive interactions between qusits and the resulting non-linear dynamics open the door of unexpectedness and creativity, as in the case of the terrestrial biosphere. In any case, we could theorize on U, we could use it as a theoretical reference to grasp the essence, magnitude and possibilities of real universes. Colossal as it may seem, U would be a finite object and then composed of a number of elements incomparably less than the number of points $(2^{\aleph_\circ})$ a simple interval of, say, one trillionth of a millimeter of the continuous space. In addition, while the points of the space continuum are abstract artifacts devoid of intrinsic physical attributes, each element of U would be plenty of intrinsic physical meaning.

Paper 7. Preinertia

Abstract.-Preinertia is an implicit assumption in classical and modern physics that is made explicit here so that it can be used in some very significant discussions. It is a universal property of all physical objects, including massless elementary particles such as photons. Indeed, it is proved here that photons are preinertial. And it is discussed on what universal property of elementary particles preinertia could be based. It is also proved that preinertia makes impossible to detect absolute motion, with the exception, perhaps, of the motion of Earth *through* the CMB isotropic frame. Preinertia will also be used in the subsequent parts of this series of articles to discuss the gravitational interaction between photons and massive bodies, suggesting a much simpler alternative to the deformation of the (inconsistent) spacetime continuum.

Keywords: preinertia, inertia, Principle of Inertia, mass, absolute motion, relativistic preinertia, CMB.

7.1 Introduction

Though only implicitly, preinertia already appears in Galileo's first relativistic discussions. Specifically in the discussion about the fall of the lead ball thrown from the top of the vertical mast of a ship moving with a uniform velocity v. In that discussion, Galileo refutes the Aristotelian conception of motion [115, p. 106-275]. Indeed, and contrarily to the hegemonic Aristotelian opinion, Galileo defended the ball hits at the base of the mast [115, p. 126-127]. According to Galileo, for the observers in the ship the ball follows a vertical trajectory while for the observers in the dock the ball follows a non-vertical trajectory.

But there is a fact that will be observed in the same way by all

91

observers, those in the ship and those in the dock: the ball always moves parallel to the mast of the ship. And here is where preinertia appears: for the observers in the ship the ball moves parallel to the mast because no force other than gravity act on the ball while falling down; this is also true for the observers in the dock, but for these observers the ball can only move parallel to the mast if it continues to move with the same relative velocity v of the mast. Or in other words, if the ball INHERITS and MAINTAINS the relative velocity v as it falls down. This inheritance is a consequence of the Principle of Inertia. A consequence that, for the reasons that we will see later, deserves to be explicitly emphasized:

Definition 7.1 (of Preinertia) *A physical object is said preinertial if it inherits the relative velocity vector of the reference frame where it is set in motion.*

In the next section it will be proved that all physical objects, including massless objects as photons, are preinertial. Preinertia is a universal property of all physical objects with the highest empirical evidence. A universal property that opens the door to some interesting discussions.

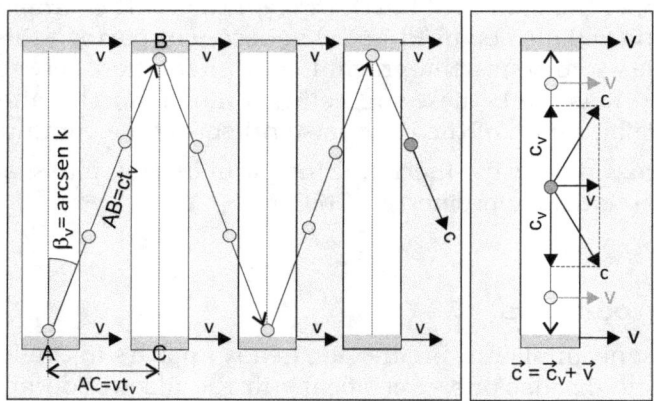

Figure 7.1 – Two ways of observing Einstein's clock of light in relative motion from the perspective of RF_v.

7.2 Photons are preinertial

(The following sections are taken from [197])
Let us consider a photon reflecting vertically on two horizontal mirrors (Einstein's clock of light) in the proper reference frame RF_o of the mirrors. In RF_v, from whose perspective RF_o moves from

left to right at a velocity $v = kc$, $(0 < k < 1)$, parallel to the direction of the increasing X_v, the photon follows a trajectory inclined with respect to the vertical by an angle β_v (Figure 7.1, left) given by:

$$\sin \beta_v = \frac{vt_v}{ct_v} = k \tag{1}$$

$$\beta_v = \arcsin k \tag{2}$$

In RF_v the motion of the reflecting photon can also be referred to the moving vertical walls of the clock, as Figure 7.1 (right) shows. This way of referring the motion of the reflecting photon illustrates that both, the clock and the photon, move with the same relative velocity $v = kc$ with respect to RF_v, the photon also moves vertically with a velocity $\vec{c}_v$ such that $\vec{c} = \vec{c}_v + \vec{v}$.

Now we will prove that photons inherit the relative velocity vector of their emitting source (as a component of its own velocity vector) whatsoever be the proper inclination at which they are emitted. For this, consider a photon a^* that is emitted at any angle α_o with respect to the X_o axis of the proper reference frame RF_o of its emitting source. Obviously, after a proper time t_o this photon will have traversed a horizontal distance d_{ox} and a vertical distance d_{oy} such that:

$$d_{ox} = t_o c \cos \alpha_o \tag{3}$$

$$d_{oy} = t_o c \sin \alpha_o \tag{4}$$

The corresponding components c_{ox}, c_{oy} of its velocity vector will be:

$$c_{ox} = c \cos \alpha_o \tag{5}$$

$$c_{oy} = c \sin \alpha_o \tag{6}$$

Assume the frame RF_v coincides with RF_o at the precise instant $t_{oo} = t_{vo} = 0$ when the photon a^* is emitted in RF_o, being t_{oo} and t_{vo} respectively measured in RF_o and RF_v. From the perspective of RF_v, the frame RF_o moves from left to right parallel to X_v at a uniform velocity $v = kc$, $(0 < k < 1)$. Thus, for the observers in RF_v the photon a^* travels a vertical distance d_{vy}:

$$d_{vy} = d_{oy} = ct_o \sin \alpha_o \tag{7}$$

in a time t_v:

$$t_v = \gamma t_o + \frac{\gamma(t_o c \cos \alpha_o)kc}{c^2} \tag{8}$$

$$= \gamma t_o(1 + k \cos \alpha_o) \tag{9}$$

Therefore, and taking into account that $\gamma^{-1} = \sqrt{1 - k^2}$, the vertical component c_{vy} of the velocity vector of the photon a^* will be:

$$c_{vy} = \frac{ct_o \sin \alpha_o}{\gamma t_o(1 + k \cos \alpha_o)} \tag{10}$$

$$= \frac{c \sin \alpha_o}{\gamma(1 + k \cos \alpha_o)} \tag{11}$$

$$= \frac{c\sqrt{1 - k^2} \sin \alpha_o}{1 + k \cos \alpha_o} \tag{12}$$

To calculate the horizontal component c_{vx} of the velocity vector of the photon a^* in RF_v we will assume the universality of the speed (modulus of the velocity vector) of light. Accordingly, we can write:

$$c_{vx}^2 = c^2 - c_{vy}^2$$

$$= c^2 - \frac{c^2(1 - k^2) \sin^2 \alpha_o}{(1 + k \cos \alpha_o)^2}$$

$$= \frac{c^2(1 + k^2 \cos^2 \alpha_o + 2k \cos \alpha_o) - c^2 \sin^2 \alpha_o + k^2 c^2 \sin^2 \alpha_o}{(1 + k \cos \alpha_o)^2}$$

$$= \frac{c^2 + c^2 k^2 \cos^2 \alpha_o + 2c^2 k \cos \alpha_o - c^2 \sin^2 \alpha_o + k^2 c^2 \sin^2 \alpha_o}{(1 + k \cos \alpha_o)^2}$$

$$= \frac{c^2(1 - \sin^2 \alpha_o) + k^2 c^2 + 2c^2 k \cos \alpha_o}{(1 + k \cos \alpha_o)^2}$$

$$= \frac{c^2(\cos^2 \alpha_o + k^2 + 2k \cos \alpha_o)}{(1 + k \cos \alpha_o)^2}$$

$$= \frac{c^2(k + \cos \alpha_o)^2}{(1 + k \cos \alpha_o)^2}$$

And then:
$$c_{vx} = \frac{c(k + \cos\alpha_o)}{1 + k\cos\alpha_o} \tag{13}$$

A little algebra suffices now to prove that the horizontal component c_{vx} of the photon velocity with respect to RF_v given by (13) implies that our photon a^* inherited (in vector terms) the relative velocity kc of its emitting source (and then of RF_o) with respect to RF_v:

$$c_{vx} = \frac{c(k + \cos\alpha_o)}{1 + k\cos\alpha_o} \tag{14}$$

$$= \frac{kc + c\cos\alpha_o + k^2 c\cos\alpha_o - k^2 c\cos\alpha_o}{1 + k\cos\alpha_o} \tag{15}$$

$$= \frac{(1 - k^2)c\cos\alpha_o + kc + k^2 c\cos\alpha_o}{1 + k\cos\alpha_o} \tag{16}$$

$$= \frac{(1 - k^2)c\cos\alpha_o}{1 + k\cos\alpha_o} + kc \tag{17}$$

$$= \frac{(1 - k^2)^{1/2}c\cos\alpha_o}{(1 - k^2)^{-1/2}(1 + k\cos\alpha_o)} + kc \tag{18}$$

$$= \frac{\gamma^{-1}c\cos\alpha_o}{\gamma(1 + k\cos\alpha_o)} + kc \tag{19}$$

$$= \frac{\gamma^{-1}t_o c\cos\alpha_o}{\gamma(t_o + kt_o\cos\alpha_o)} + kc \tag{20}$$

$$= \frac{\gamma^{-1}t_o c\cos\alpha_o}{\gamma t_o + \dfrac{\gamma kt_o c^2\cos\alpha_o}{c^2}} + kc \tag{21}$$

$$= \frac{\gamma^{-1}t_o c\cos\alpha_o}{\gamma t_o + \dfrac{\gamma(t_o c\cos\alpha_o)kc}{c^2}} + kc \tag{22}$$

$$= \frac{\gamma^{-1}t_o c\cos\alpha_o}{t_v} + kc \tag{23}$$

$$= \frac{\gamma^{-1}t_o c \cos \alpha_o + kct_v}{t_v} \tag{24}$$

Therefore, during the time t_v, and with respect to RF_v, the photon a^* runs through a horizontal distance d_{vx}:

$$d_{vx} = \gamma^{-1}t_o c \cos \alpha_o + kct_v \tag{25}$$

The right side of (25) has two terms:

1) According to (3), the first term $\gamma^{-1}t_o c \cos \alpha_o$ is the horizontal distance a^* moves with respect to RF_o for the time t_o, although contracted by the relativistic factor γ^{-1}. This would be the horizontal distance our photon a^* would have traversed with respect to RF_v if it had not inherited the relative velocity of its emitting source.

2) The second factor kct_v is the distance the emitting source moves with respect to RF_v during the time t_v with a velocity kc.

all of which suggests that the velocity vector of the photon a^* inherited the relative velocity vector as part of its component parallel to the direction of relative motion. And indeed, equation (25) allows us to prove that this is the case:

$$c_{vx} = \frac{d_{vx}}{t_v} \tag{26}$$

$$= \frac{\gamma^{-1}t_o c \cos \alpha_o + kct_v}{t_v} \tag{27}$$

$$= \frac{t_o}{\gamma t_v} c \cos \alpha_o + kc \tag{28}$$

$$= \frac{t_o}{\gamma^2 \left(t_o + \frac{kt_o c \cos \alpha_o}{c} \right)} c_{ox} + kc \tag{29}$$

$$= \frac{1}{\gamma^2 (1 + k \cos \alpha_o)} c_{ox} + kc \tag{30}$$

$$= \frac{1 - k^2}{1 + k \cos \alpha_o} c_{ox} + kc \tag{31}$$

This conclusion is confirmed by the following argument: Assume that with respect to the X_v axis of RF_v the photon a^* only travels the distance:

$$d_{vx} = \gamma^{-1} d_{ox} \tag{32}$$

Since $d_{ox} = t_o c \cos \alpha_o$ (3), equation (23) would give rise to:

$$c_{vx} = \frac{\gamma^{-1} t_o c \cos \alpha_o}{t_v} + kc \tag{33}$$

$$= \frac{\gamma^{-1} d_{ox}}{t_v} + kc \tag{34}$$

$$= \frac{d_{vx}}{t_v} + kc \tag{35}$$

$$= c_{vx} + kc \tag{36}$$

$$\therefore\ k = 0 \tag{37}$$

which is not the case because $k > 0$.

In consequence, once emitted, and from the perspective of RF_v, our photon a^* moves in the direction of the relative motion (apart from the projection of its inclined trajectory on that direction) the same distance and for the same time as its emitting source (25), and it inherits the relative velocity vector as a part of its vector component parallel to the direction of the relative motion (31). The other part of this component is a fraction of c_{ox} defined by the complementarity factor f_{vx}, which according to (31) is:

$$f_{vx} = \frac{1 - k^2}{1 + k \cos \alpha_o};\ 0 \leq f \leq 1 \tag{38}$$

which, obviously, decreases with the relative velocity kc and increases with α_o (Figure 7.2). The other vector component c_{vy} will be defined according to:

$$c^2 = c_{vx}^2 + c_{vy}^2 \tag{39}$$

$$c_{vy}^2 = c^2 - c_{vx}^2 \tag{40}$$

$$= c^2 (1 - f_{vx}^2 \cos^2 \alpha_o) \tag{41}$$

It is then clear that from the perspective of the reference frame RF_v, the velocity vector of the photon a^* inherits as a part of one of its components the relative velocity vector of its emitting source. At this point, and according to all the theoretical and experimental evidence, we have no choice but to accept that photons are preinertial; that they, in fact, inherit the relative velocity vectors of their emitting sources as a component of their own velocity vectors (except, perhaps, in the case in which both vectors are parallel, which would be a sort of test on the prevalence of preinertia on the universality of the speed of light, or vice versa).

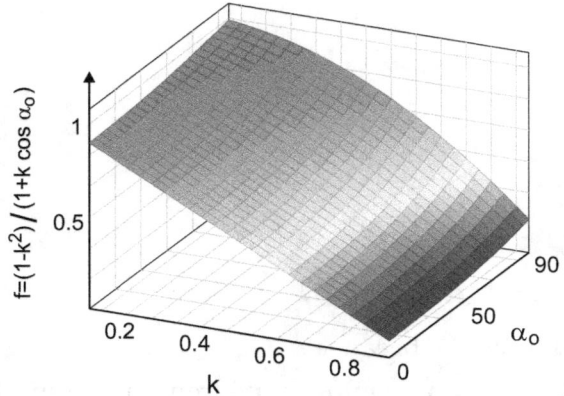

Figure 7.2 – Complementarity factor $f = (1 - k^2)/(1 + k\cos\alpha_o)$ in terms of k and α_o.

We must conclude that whenever the direction of the relative velocity v of a source of photons is different of the emitting direction d_e, all emitted photons inherit the relative velocity vector $\vec{v}$ of the source as a component (or a part of a component) vector of its own velocity vector, the other component in the plane defined by v and d_e being such that the module of the resulting vector is the universal speed of light c.

The above conclusion, applies to all directions, except to the direction of the relative motion of its emitting source. This exception has an immediate (and axiomatic) explanation: the Second Principle of Relativity. If the photon is emitted in the same direction and sense as the relative motion, then the exception could also have a physical explanation in a discrete space and time: the existence of a maximum insurmountable velocity of one unit of space per one unit of time. If the photon is fired in the same direction as the relative motion but in the opposite sense, then only the axiomatic explanation remains. So, for all directions and senses the inherited relative velocity has a physical explanation, except in the case just indicated, which is the case that motivates Santiago del Collado experiment [197, pp. 463-488]. Obviously, the above argument on photons can also be applied to any other physical object, be it, or not, an elementary particle, and be it or not a massive object. It, then, proves the following:

Theorem 7.1 (of Preinertia) *Every physical object inherits in one of its vector components the relative velocity vector of the reference frame where it is set in motion, provided that the resulting speed does not exceed the possible maximum limit.*

We have made use of Lorentz Transformation, and therefore of the special relativity, to demonstrate the preinertial nature of photons. So it can be said that special relativity implies preinertia. But preinertia being such a basic and universal property of matter, and there being so much empirical evidence confirming it (for example, each time an object falls to the ground), we may wonder whether the principles of relativity are necessary to demonstrate preinertia, or is preinertia an aspect of the Principle of Inertia that has gone unnoticed, perhaps because of its excessive evidence. This enormous empirical evidence of preinertia recommends making it independent of special relativity (SR) and incorporating it into the statement of the Principle of Inertia by simply adding three words, namely:

IS, PREINERTIAL, AND.

Thus, in order to make preinertia explicit, the Principle of Inertia could be stated as follows:

Principle 7.1 (of Inertia) *Every physical object is preinertial and remains at rest or moves at a constant uniform velocity, unless an external force acts upon it.*

We accept the veracity of the Principle of Inertia because of its extraordinary empirical evidence. It is an inductive principle that we use, along with others, to begin to construct an explanation of the physical world, the construction of physics. Now, should that be the starting point, or can we try to solve some even more basic questions? For example:

Let A and B be any two physical object of the same type at rest in their common inertial reference frame RF_o. Let now A be set in linear uniform motion with respect to RF_o.

- What determines and controls the linear trajectory of A, its successive positions along the successive instants?

- How does A remember that it was set in motion? Where lies the imprint of that action?

- What changed, if any, in the internal structure of A as a consequence of being set in motion?

- What distinguishes an object that has been set in motion from another that was not?

- Is space and time somehow affected by an object set in motion?

- Knowing that A was set in motion and B was not, is it the same to say that A moves with respect to B as to say that B moves with respect to A?

- If A is a massless photon, what quality of its nature determines its preinertia? Or is it not a massless object?

- To set A in motion is the same as to set the rest of the universe in motion?

- Is there any absolute describable reality?

- If there is no reality describable in absolute terms, are there as many realities as there are relative forms of observing it? To observe what?

- Could the universe be described, as such an object, from outside the universe?

- Are we living beings endowed with the capacity to reason but not to observe reality?

- Is the theory of special relativity the ultimate theory?

- etc.

As Feynman said, we know how objects move but not why they do (why they move in a straight line) [105, p. 18]. But should science give up answering the above questions?

7.3 Preinertia and absolute motion

The impossibility to measure absolute velocities has been confirmed in experimental terms, but its derivation from the principles of relativity is axiomatic, and then empty of empirical meaning. The impossibility to measure absolute velocities is better explained in physical terms by the preinertial nature of photons and, according to the Principle 7.1 of Inertia, of all physical objects. There would be a sort of mechanical entanglement between all physical objects of an inertial reference frame (even if they are created in that frame). And the entanglement is maintained forever, unless a force modifies it. As will be seen below, this mechanical entanglement of all objects in an inertial reference frame makes it impossible to use just these objects, once set in motion, to detect the absolute motion of the frame were they are set in motion. Although, as indicated above, there could be an exception.

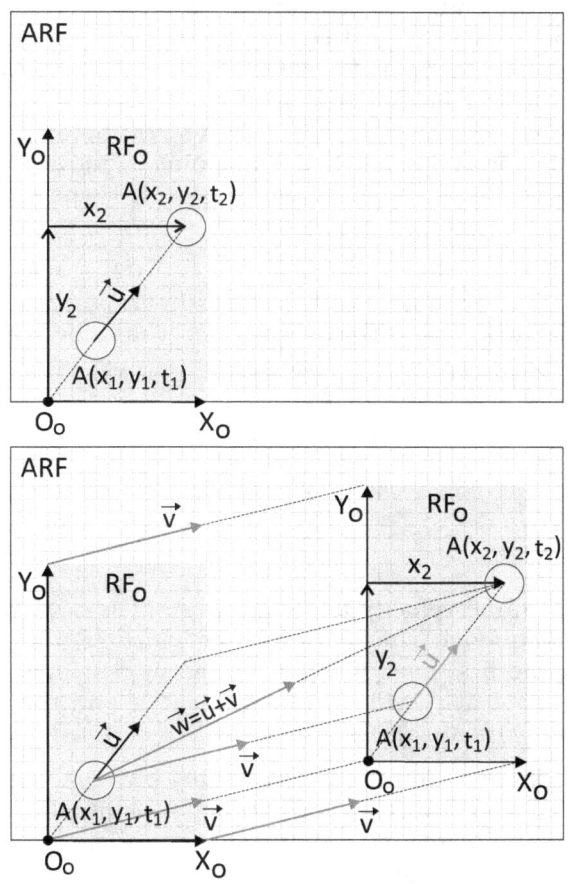

Figure 7.3 – Absolute motion is undetectable. Top: the physical object A is set in motion in its proper reference frame RF_o at rest in RF_a. Bottom: the physical object A is set in motion in its proper reference frame RF_o, which moves through RF_a with an absolute velocity $\vec{v}$.

Assume, just for a moment! that there exist an absolute reference frame RF_a (perhaps made of the indivisible quantum space units we have termed here qusits) through which physical objects can move in absolute terms. Let RF_o be a reference frame at rest in RF_a, and let A be any physical (point) object at rest in RF_o, where it is placed in the position (x_1, y_1, t_o) of RF_o (for simplicity, we dispense with the z-coordinate). Let A be set in motion at t_1, $(t_o < t_1)$ with a uniform velocity $\vec{u}$ so that at the instant t_2 it is placed in the position of coordinates (x_2, y_2, t_2) (see Figure 7.3, top).

Consider now that RF_o moves in RF_a with an absolute and uniform velocity $\vec{v}$, and let A be set in motion under the same above conditions when RF_o was at rest in RF_a. Thanks to preinertia, A *inherits* the absolute velocity vector $\vec{v}$ of RF_o with respect to RF_a, and thanks to the Principle of Inertia, A *maintains* $\vec{v}$ along its own motion with respect to RF_o (Figure 7.3, bottom). Let O_o be the origin of coordinates of RF_o. This point O_o moves with respect to RF_a at a velocity $\vec{v}$, while A moves with respect to RF_a at a velocity:

$$\vec{w} = \vec{u} + \vec{v} \tag{42}$$

The object A (that moves with respect to RF_a at the velocity $\vec{w}$ given by (42)) will move with respect to O_o (that moves with respect to RF_a at the velocity $\vec{v}$) at a velocity $\vec{u'}$ given by:

$$\vec{u'} = \vec{w} - \vec{v} \tag{43}$$

$$= \vec{u} + \vec{v} - \vec{v} \tag{44}$$

$$= \vec{u} \tag{45}$$

which is the same velocity as if RF_o were at rest with respect to RF_a. In consequence, the coordinates of A in RF_o at t_2 will be the same as in the first case when RF_o was at rest in RF_a. So, the coordinates of A at t_2 will also be (x_2, y_2, t_2), and they cannot be used to detect the absolute motion of RF_o.

Since RF_o is any reference frame, A any physical object initially at rest in RF_o, and $\vec{u}$ any uniform velocity, we must conclude that the absolute motion of a reference frame is undetectable by setting into motion any physical object (or objects) of that reference frame.

The only way to eliminate the inherited velocity of an object set in motion in a reference frame would be to apply to it a force capable of eliminating that velocity, for which that velocity would first have to be known. It does not seem possible, then, to know the absolute velocity of a reference frame with the only help of the objects of that reference frame, or created in that reference frame. Experiments a la Michelson Morley would always give, for this reason, negative results. Let me now recall again with admiration Newton's words [240, Corollary V, p 144]:

The motions of bodies included in a given space are the

same among themselves, whether that space is at rest, or moves uniformly forwards in a right line without any circular motion.

The above argument also applies to the case of elementary particles set in motion in the same conditions as A (or even created and set in motion in those conditions). It could be argued, however, that the argument only applies to elementary particles as such particles, but not to their corresponding associated waves. Evidently, if this were the case, particle-wave decoupling would occur, which is unknown in modern physics (as far as I know). It seems reasonable to conclude that positive results should not be expected in experiments a la Michelson-Morley (except, maybe, in the case of Santiago del Collado experiment above indicated).

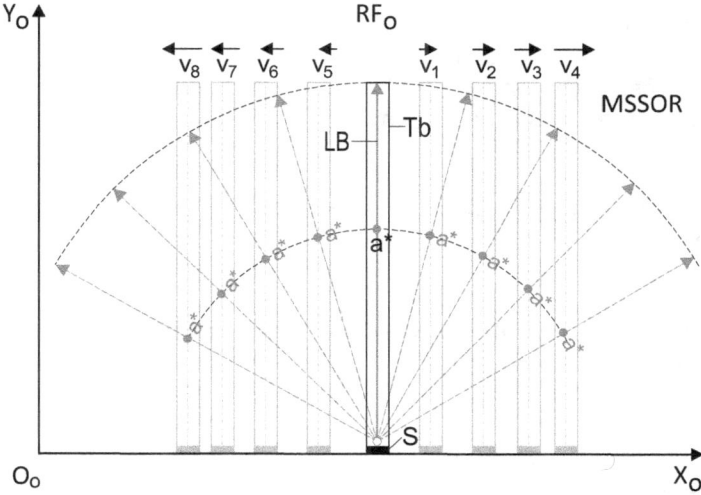

Figure 7.4 – The vertical trajectory of the photon in the proper reference frame RF_o of its source S is seen at different inclinations depending on the relative velocity at which it is observed. MSSOR: multiple superposed and simultaneous objective realities.

7.4 Two key questions

There is enormous empirical evidence that preinertia is a universal property of all physical objects, even massless physical objects such as photons. We can then ask ourselves about the reason for preinertia. If photons have spin 1 but no electric charge, no color charge and no mass, what essential property of photons (and of all other particles, since they are all preinertial) is responsible for their preinertia? This is our first key question.

To pose our second key question, consider a photon a^* emitted parallel to the Y_o axis of the proper reference frame RF_o of its source S (Figure 7.4). According to SR (and preinertia), the trajectory of the photon a^* will be different in different references frames, depending on their relative velocities respect to RF_o. According to SR, each of these trajectories is as real as any other. Recall Einstein's own words:

> ...and shows that the clock goes slower than if it were at rest relatively to K'. These two consequences, which hold, mutatis mutandis, for every system of reference, ...[81, p. 38].
> ...From all of these considerations, space and time data have a real, and not a mere fictitious, significance; [81, p. 30].
> ...It is clear that the same results hold good of bodies at rest in the "stationary" system, viewed from a system in uniform motion.[83, p. 49]

So, preinertia and SR make it inevitable to face the following two alternatives:

a) There are multiple superposed and simultaneous objectives realities (MSSOR), as many of them as possible relative velocities at which any physical object or event can be observed.

b) There is a unique objective reality in which all objects moves through the same absolute space frame.

The second alternative seems much more simple, and it is not incompatible with relative motion: relative motion is an inevitable consequence of moving with different absolute velocities with respect to the absolute space frame. The problem is that, until now, absolute motion is undetectable. And for the reasons given above, maybe it will always remain undetectable. According to this alternative, SR is a mathematical theory on apparent realities, as apparent as the deformed rod partially submerged in water. Because, in fact, the rod is not deformed no matter how many experimental measurements we make checking its deformation.

7.5 Preinertia and the nature of light

Almost every general physics textbook includes a chapter on the nature of light [320, 321, 106, 332, 297] etc., not to say publications specifically devoted to the matter, for instance [214, 52, 281, 273, 308, 104, 149, 92, 226] etc. In page 291 of the Oxford Dictionary of Physics [64] we can read (the italic is mine):

> **light** The form of electromagnetic radiation to which the human eye is sensitive and on which our visual awareness of the universe and its content relies.
>
> ...In 1905 Einstein showed that the photoelectric effect could only be explained on the *assumption* that light consists of a stream of discrete photons of electromagnetic energy.
>
> ...While it is not easy to construct a model that has both wave and particle characteristic, it is *accepted*, according to the Principle of Complementarity proposed by Niels Bohr, that in some experiments light will *appear* wavelike, while in other it will *appear* to be corpuscular.

Assumption, accepted, appear,...it seems that, in the end, we don't know what light really is. Indeed, let us see how quickly the answers to successive pertinent questions are exhausted:

1) What is light? Answer: A set of electromagnetic waves.

2) What is an electromagnetic wave? Answer: An oscillation of the electric field associated with an orthogonal oscillation of the magnetic field.

3) What is a field? Answer: A region of space in which certain forces manifest.

4) What is space? Answer:

5) What is a force? Answer:

And taking into account the preinertial nature of light, does preinertia intervene in the gravitational interactions of photons with massive objects? Could these hypothetical interactions (forces) explain the observed gravitational curvature of space without having to curve space? And as noted above, if it is not mass, what basic attribute of the nature of light could be responsible for its preinertia?

We can describe light and its propagation through different media to a certain level of essentiality. And for the same reasons as with the Principle of Inertia, we should not stop at that level of essentiality. But to continue asking and resolving more essential questions about the nature of light, we would surely have to first clarify the nature of space and time. Which, in turn, will not be possible if we continue to ignore the inconsistent nature of the actual infinity and, therefore, the inconsistency of the spacetime continuum.

Many things will change when we accept, and prove, the discrete nature of space and time. Perhaps the CALM models (see Paper 6) would be of great help in initiating this new physical exploration of the physical world. I suspect that the Planck-scale visual perception of how my office space fills with light when I turn on the lamp over my desk would be very revealing.

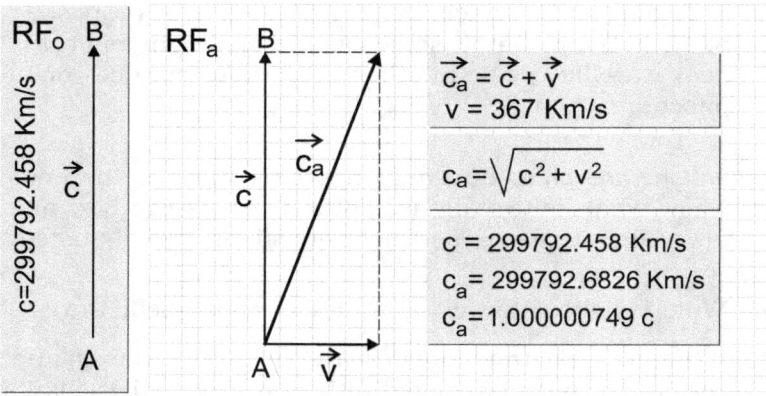

Figure 7.5 – Speed of light c and absolute speed of light c_a. The velocity $v = 367 Km/s$ is the velocity of the Earth through RF_a, the reference frame of the isotropic cosmic microwave background, assumed as an absolute reference frame.

7.6 The speed of light and absolute motion

The speed of light has been measured on different occasions, in different places and with different procedures. Today it is unanimously accepted that this speed is 299792.458 km/s. Now, since photons are of a preinertial nature, it makes sense to ask about the possible consequences on their velocity of inheriting the relative velocity vector of their corresponding emitting sources. In this sense, assume we measure the speed of light in our terrestrial reference frame (Figure 7.5, left). We must consider the following two alternatives:

a) Light inherits simultaneously a potentially infinite number of relative velocities, as many as relative velocities at which it can be observed from any possible inertial reference frame.

b) Light inherits a unique absolute velocity, which is the absolute velocity of the reference frame where it is created and set in motion.

The first alternative is very uncomfortable from the physical point of view. According to the second alternative, the measured velocity c of light in our planet would include the absolute velocity v of our planet through the same absolute reference frame RF_a (Figure 7.5, right). So, we could write for the absolute velocity of light c_a:

$$\vec{c_a} = \vec{c} + \vec{v} \tag{46}$$

$$c_a = \sqrt{c^2 + v^2} \tag{47}$$

$$= \sqrt{299792.458^2 + 367^2} \tag{48}$$

$$= 299792.6826 Km/s \tag{49}$$

Therefore, the absolute speed of light c_a would be 1.000000749 times greater than the velocity c measured here, in the Earth. The problem is that this conclusion could be impossible to confirm here in the Earth. Except, may be through the above mentioned Santiago del Collado experiment.

Paper 8. Infinite regress

Abstract.-Paper 8 of this series of papers on discrete cosmology examines the Aristotelian infinite regress of arguments, and extends it to definitions and causes. Infinite regress then becomes a very serious limitation of human knowledge, much more serious than Gödel's famous incompleteness theorems. But at the same time, and also considering the inconsistency of the actual infinity and the infinite divisibility, it suggests the way we should go to explain the physical world by clearly setting out what can and cannot be explained in formal terms. In particular, it is proved here that any history either has an initial instant or it is inconsistent.

Keywords: infinite regress of arguments, infinite regress of definitions, infinite regress of causes, axioms, principles, inductive laws, primitive concepts, origins, Münchhausen Trilemma, Agrippan Trilemma, Theorem of Formal Dependence, Theorem of the First Element.

8.1 Introduction

Although an infinite regress is any infinite sequence of elements that are recursively related (each element of the sequence is related in a certain way to its immediate predecessor in the sequence), we are interested here only in the particular cases of demonstrations, definitions, and causes. The case of the demonstrations was treated by Aristotle [13, I.3] and it will be the one with which we begin our discussion, also making use of the famous Münchhausen trilemma. The discussion will then be extended to definitions and causes.

Making use of the inconsistency of the actual infinity, the need

109

for primitive concepts, axioms, fundamental laws, and inductive principles will be demonstrated. The Theorem of the First Element will also be proved. All these results will be relevant to define the foundations of the finitist and discrete cosmology proposed in this series of articles.

The common thread of all these demonstrations will be the Theorem 5.6 of Formal Dependence, directly deduced from the Principle 5.1 of Directional Evolution. That theorem establishes that statements do not prove themselves; concepts do not define themselves; and objects are not the cause of themselves. Obviously, the alternative would be an inevitable collection of nonsense incompatible with science.

As indicated above, the infinite regress of arguments was discovered more than twenty two centuries ago. And, for the reasons that will be discussed below, it is a serious limitation of human knowledge. However, most contemporary scientists ignore it. The curious thing is that at the same time that modern science ignores these logically inevitable restrictions, it pays an exaggerated attention to other much less general limitations, almost always related to the conflicting self-reference [194].

In the particular case of physicists (experimental and theoretical) it is rare to see that they are concerned with these formal limitations, as if these limitations did not limit anything. This is not the best attitude because until the problems posed by such limitations are resolved, physics cannot be built on adequate foundations. Experimental data can force the adjustment of physical theories, but they have little bearing on their foundations. That could be the reason of the disagreement, even the incompatibility, of some of modern physical theories with each other.

8.2 Münchhausen Theorem

Also known as Agrippan Trilemma, the well-known Münchhausen Trilemma is an argument that tries to demonstrate the impossibility of proving the truth of any statement without making use of an arbitrary initial statement. According to this argument, a truth can only be proved by means of:

1. An infinite regress of proofs.

2. A first arbitrary statement.

3. A circular sequence of proofs.

The three of which are formally unsatisfactory. As we will see now, things do not improve with the Theorem 5.6 of Formal Dependence and the Theorem of the First Element 8.2. Arguably they get worse because they become a theorem whose consequences are the same as the Münchhausen Trilemma. Here, we will deal only with infinite regress of proofs, definitions, and causes, in relation to which it is immediate to demonstrate the following:

Theorem 8.1 (of the Uncompletable Regress) *Any recursive sequence S of proofs, definitions or causes in which there is a last element to be proved (defined, caused) and each element has an immediate predecessor that proves (defines or causes) it, is uncompletable.*

Proof: If every element of the sequence has an immediate predecessor, then there is not a first element of the sequence, because this first element would have no immediate predecessor. Therefore, the sequence, if consistent, can only be potentially infinite and uncompletable (Corollary 3.5). □

In addition, the third option of the Münchhausen Trilemma would be inconsistent because there would be at least one element that proves (defines or is the cause) of itself, which goes against the Theorem of Formal Dependence 5.6.

8.3 A significant theorem

As indicated in the introduction, an infinite regress is a sequence of recursively related elements such that each element maintains the same type of relationship with its immediate predecessor. In our case the elements of the sequence will always be formal elements: proofs, definitions and causes. And their relationships will always be the same formal relationship: proofs in the infinite regress of proofs; definitions in the infinite regress of definitions; and causes in the infinite regress of causes.

We will now prove a general result that applies immediately to infinite regress:

Theorem 8.2 (of the First Element) *A consistent sequence in which there is a last element and each element has an immediate predecessor is a complete totality only if it has a first element.*

Proof: Let $S = \ldots S_{3*}, S_{2*}, S_{1*}$ be any sequence with a last element S_{1*} and in which each element S_{n*} has an immediate predecessor $S_{(n+1)*}$, where $n*$ read last but $n - 1$. If S is consistent it can only be finite or potentially infinite (Corollary 3.5). Therefore, if S is a complete totality it can only have a finite number n of elements

(Definition 3.5, page 32). In these conditions, and taking into account that each element S_{i*} of S has exactly one predecessor more than its immediate predecessor $S_{(i+1)*}$, the element S_{1*} has $n-1$, predecessors; the element S_{2*} has $n-2$ predecessors; the element S_{3*} has $n-3$ predecessors; etc. Consequently, the smallest number of predecessors that an element of S can have is $n-(n-1)=1$. That element will be $S_{(n-1)*}$ whose predecessor can only be a first element S_{n*} of the sequence that has no predecessor. So, S has a first element S_{n*} with zero predecessors. $\square$

8.4 Infinite regress of proofs

An immediate, and well known, consequence of Theorems 8.1 and 8.2 is that all formal sciences must be founded on a set of axioms, i.e. a set of statements whose veracity is assumed without proof (see next section for the role of definitions and primitive concepts in the foundations of formal sciences). Ideally, the number of axioms should be small and as self-evident as possible (otherwise, they would give rise to an excessively arbitrary science).

As is well known, classical Euclidean geometry was founded on five axioms, the fifth of which is the controversial axiom of parallels, whose statement is anything but self-evident. By contrast, Playfair and Hilbert founded their Euclidean geometries respectively on 30 and 20 axioms. As expected, the initial set of axioms will have important consequences on the resulting of science [193, 257, 144]. For example, Euclidean and non-Euclidean geometries.

For its part, set theory is based on about ten axioms (depending on the version). One of them is the Axiom of Infinity. Assuming this axiom has enormous consequences, not only in mathematics but also, and above all, in physics, a science that pays little attention to the foundations of its mathematical language, and considers that this language is not an instrument for analyzing observations but a model to which observations must be adapted.

Since the beginning of the 20th century, physics has been built up with a type of mathematics (the infinitist mathematics of the spacetime continuum) that assumes the existence of the complete list of the natural numbers in their natural order of precedence (which is a way of stating the Axiom of Infinity), even though there is no last natural number to complete the list. They assume, as

Aristotle would say, that the incompletable exists as complete.

The formal proofs of the inconsistency of the Axiom of Infinity has been available for more than twenty years, but it is taking too much effort to fight against the infinitist stream that assumes that axiom. An stream of thought absolutely dominant and hostile to dissidence. But in the end, well-constructed proofs will end up imposing their conclusions. The reader will be able to judge one of those proofs in paper 3 (Theorem 3.5 of the Denumerable Infinity) in this series, and forty others in [198, pdf link]. The consequences on physics will be enormous, and very positive. This series of articles has a lot to do with that finitist and discrete future. Indeed, the inconsistency of the actual infinity will change all.

Experimental sciences have an additional element for their corresponding foundations: the inductive principles or laws. Statements whose veracity is accepted without formal proofs (such as axioms) but confirmed by experimentation and observation of natural phenomena. This is the case, for example, of the Principle of Inertia, including preinertia (see Paper 7). But here, too, it is important to be careful. In this sense, it is convenient to recall Russell's famous metaphor of the chicks [285, p. 31]): the innocent animals who lived happily on the farm in the care of their attentive farmer without suspecting the existence of fried chicken with potatoes. Fortunately, humans are, in general, smarter than chicks and we have discovered that it is wise to be cautious when drawing conclusions about the physical world.

8.5 Infinite regress of definitions

Although the infinite regress has always been discussed for the case of demonstrations, its extension to other formal elements such as definitions and causes is immediate. And the consequences of Principle 5.6 and Theorem 8.2 on these new formal elements are also immediate.

Since concepts are not self-defining (Theorem 5.6), to define any concept it is necessary to use one or more different concepts; and the same goes for the latter. In this way a recursive sequence of definitions appears to which Theorem 8.2 can be applied, with the same consequences as in the case of axioms:

Corollary 8.1 (of Primitive Concepts) *Primitive (undefinable) concepts are inevitable in all sciences and languages.*

Proof: It is an immediate consequence of Theorems 5.6 and 8.2.
□

Naturally the defined objects, and the undefined ones, have to be legitimated by the axioms or principles of the corresponding theories, or by formal proofs. This is usually the case in the formal sciences, but not always in the experimental sciences. Even in theoretical physics or theoretical biology, which are more formal and mathematical than their corresponding experimental branches. The problem is that when a science pay little attention to its fundamental base anything can happen (see paper 2).

The most basic concepts of science such as set, number, point, force, mass, time etc. are primitive concepts. Most are intuitive: We know what they are even though we have no formal definition of them. This is the case, for example, of set, number, force or time. In other cases we may have a wrong intuition. In my opinion that's the case of point and instant. The intuition we have of point is confused with that of the mark on the paper or board that tries represent it. And the intuition of instant is confused with very short intervals of time (paper 11 deals with these issues).

The concept of point is primitive and fundamental in physics: space would be made up of points without extension, points that do not occupy places but define places; and there would be point masses, point charges, point particles, point trajectories, etc. And if that was not enough, there are the same number of points on a line one trillionth of a millimeter long as there are in the entire three-dimensional universe. Obviously, the same could be said of the concept of instant: the same number of them elapses in a micro-second as in the whole history of the universe. This is Cantorian infinitism!

Appropriate efforts may not always have been made to define objects and to establish the axioms and proofs that legitimize them. Naturally, definitions have to be formally productive: they have to be usable in subsequent proofs. A very notorious case is that of Euclidean geometry: it is possible to define a new foundational basis with 28 productive definitions and 10 axioms in which it is possible to prove, like any other theorem, the statement of the Euclidean axiom of parallels [193].

It is worth clarifying this issue further. The concepts of point, line and straight line are primitive. So a straight line (a central concept in Euclidean geometry) is something for which we do not have a formal definition; in turn a straight line belongs to a class of objects (lines) for which we do not have a definition either; and a line is made up of points, of which we do not have a definition either. Perhaps too many 'indefinitions'.

It is possible, however, to give a formal definition of a straight line (although the concepts of point and line remain primitive). It is a formally very productive definition that, together with the rest of the new foundational elements of Euclidean geometry, allows us to demonstrate the Euclidean statement of parallels [193, p. 40].

8.6 Infinite regress of causes

In accordance with Theorem 5.6 of Formal Dependence, physical (and formal) objects are not self-causing, i.e. if they cannot be the cause of themselves. In consequence Theorems 8.2 and 8.1 apply to them. We must, therefore, accept the following:

Corollary 8.2 (of the First Cause) *To explain the origin of any physical object or phenomenon, there must be a first cause that cannot be explained in terms of other causes derived from our knowledge of the observable universe.*

Proof: It is an immediate consequence of Theorems 5.6 and 8.2. □

Paper 15 deals with the corresponding consequences of Corollary 8.2, which, as the reader can imagine, will be anything but irrelevant. In the meantime, do not forget that science should be free of prejudices, free even of religious and anti-religious prejudices. And that, on the other hand, without the necessary formal rigor, language cannot be scientific. Without the rigorous use of language, anything can be demonstrated. To affirm, for example, that the universe arose from a fluctuation of nothingness, implies that nothingness is not nothingness but something with the capacity to fluctuate universes. As T. Maudlin would surely say, the importance of the following conclusion of Corollary 8.2 of the First Cause cannot be exaggerated: The evolution of the universe, as such a natural process, must also have a first cause outside the evolution of the universe itself; a first cause that cannot be explained in terms of other causes deduced from our knowledge of the observable universe, i.e. a first cause that cannot be explained in physical or logical terms.

Paper 9. Discrete space

Abstract.-This paper discusses the consequences (on the physical space) of some of the formal results proved in the precedent papers, namely the Theorem 5.6 of Formal Dependence and the inconsistencies of the actual infinity and the infinite divisibility. The consequences are almost immediate and very clear, space would have to be finite, discrete and physical, i.e. real, which obviously contrast with the usual concept of space in contemporary physics. A model based on Cellular Automata Like Models is then proposed to start the discussions on the foundational basis of a new finitist and discrete paradigm of physical space.

Keywords: relative space, absolute space, finite space, infinite space, discrete space, physical space, CALM.

9.1 Introduction

Space is one of those concepts that have always attracted the attention of philosophers and physicists (and in this case also of theologians). Discussions about space began with the pre-Socratics and have continued to this day. And they have basically focused on three aspects of space: its real or fictitious nature; its absolute or relative nature; and its finite or infinite size. Leaving aside the *extensive* points of the early Pythagoreans and the discrete physical world proposed by the authors of the Islamic culture Kalam (or Kalām) (see paper 4), the discussion of the continuous or discrete nature of space is much more recent and reduced to a minority of authors.

The aim of this article is not to review the historical details of those discussions, for which there is an excellent literature [117, 338, 279, 132, 204, 213, 269, 216, 217, 165, 153, 335,

117

174, 175], but to introduce these disputed aspects of space and subsequently analyze them in the light of Theorem 5.2 on Finite Distances and Theorem 5.12 of Adjacency, both proved in paper 5.

Once the classical alternatives have been introduced, this paper discusses the formalism of one of them: the hegemonic spacetime continuum. Following this discussion, some problems are raised about the contradictory use of certain concepts related to the content of space and the relationships between the elements proper to space and the objects contained in space. These problems are rarely analyzed, though in my opinion are necessary in order to understand the role of space in the physical world.

In agreement with the formal results proved and assumed in the precedent papers (inconsistency of the actual infinity, inconsistency of the infinite division and Corollary 5.8 of Discrete Threshold), this paper 9 proves some important results on the physical nature of space, which, if nature is consistent, can only be finite, discrete and physical. Where physical means real, within the real versus relative debate.

An outline is then presented of what could be a discrete and absolute physical space and its relation with physical objects, a relation that now would not be that of a simple container but that of a generator of all physical objects. In these conditions, the concept of CALM (Cellular Automata Like Model) is proposed as an inspiring instrument to begin the construction of a new finitist and discrete paradigm of space and time, and then of the universe.

One of the obvious advantages of the new discrete paradigm is its complete independence of the Axiom of Infinity. In addition, the above mentioned relationships between the elements of space and the elements of the physical objects contained in physical space are better explained in this new discrete and finitist paradigm of physical space.

Although discrete geometries (and discrete mathematics) exist, they are not the ones that would have to be developed for this new paradigm. As an informal example (informal, because it is not built on a defined foundational base), a discrete version of Pythagoras theorem will be introduced in paper 14. This example is interesting because of the role played by this theorem in the calculation of distances, and because the factor that converts between its classical and discrete versions is the relativistic Lorentz factor γ given by:

$$\gamma = 1/\sqrt{1 - v^2/c^2} = (1 - k^2)^{-1/2}; \ 0 < k < 1 \qquad (1)$$

where, as is well known, $v = kc$ is the relative velocity and c the speed of light.

9.2 Absolute (real) and relational (fictitious) space

Is space real, independent of the existence of observable physical objects? or is it purely relational, i.e. a mental construct of our way of perceiving the physical world? This is the classical debate on the nature of space, (substantival versus relational alternatives). Or in other words, while for some authors space is a real physical entity, for others it is just a fiction, a mental way of representing objects and their relative positions. For the first ones there would be space even if it does not contain objects, and for the others there is no space if there are no objects that define it (in theoretical terms).

Practically all the authors interested in the nature of space supported one of the two alternatives, giving rise to certain discussions that are still remembered, such as that of Newton-Clarke (absolute space) versus Leibniz (relational space). The debate is still open today, although according to the hegemonic theories of relativity (special and general) space and time are only relative while spacetime is absolute, the three of them being (for most authors since Leibniz) theoretical constructs that are necessary to describe the world, but not real entities.

The debate on the nature of space is far from closed, because, in spite of being unreal, it can deform, expand and vibrate (some space deformations propagate through space itself: gravitational waves). We would have to admit the existence of an object that is not real but that can deform, vibrate and is the medium of propagation of its own vibrations. What kind of non-reality would that be? A vibration of something that does not exist, does not exist either.

On the other hand, the space that expands in the universe is not the intra-galactic space but the intergalactic space of the emptiest zones of the universe. So it is space, and only space, which expands in an accelerated way. It is also difficult to assume that an unreal object can expand, however the unreal space expands, and then self-creates continuously. Therefore, space is the cause of itself, a conclusion that goes against the Theorem 5.6 of Formal Dependence.

In addition, space would be filled by fields of the four forces and by a special field that would fill the entire spacetime: the quantum vacuum field, from which continuously emerges a multitude of virtual particles that only exist for extremely short intervals of time. The debate is further complicated by the imprecise use of certain key terms such as *nothing*, *void*, *infinity*, and even *point* and *instant*.

9.3 Finite and infinite space

The other classic debate about the nature of space is whether its extent is finite or infinite. Almost invariably, in these discussions the concept of infinity is used in a very imprecise way. And the imprecision was justified as long as no proper formal definition of infinity was available, basically until the early 20th century, when Dedekind's Definition 3.3 and the Axiom of Infinity 3.1 were proposed and accepted.

Since then, it is not justified to use the concept of infinity in imprecise terms. Despite this, it is still used in these imprecise Gaussian terms (infinity as a way of speaking, see paper 3), very common in contemporary physics, especially in the secondary literature (though not uncommon in the primary literature). The same thing happens with other concepts such as *point* or *nothingness* or *void*, even with concepts alien to our discussion but very important in physics, such as *order*, *organization* and *complexity*. As noted in Paper 2, it could be said that contemporary physics lacks some formal rigor in the use of ordinary language and formal language.

There is a third alternative regarding the finite/infinite extent of physical space. This third alternative proposes a finite but unbounded space, as with the curved surface of any sphere. In this case, the universe would have a finite size, but somehow it would be curved back on itself so that a straight trajectory would always end up at the initial point of that same trajectory. This alternative has, therefore, certain requirements on the geometry of the space itself (to be discussed in paper 11).

In papers 3, 4, 5 of this series of papers, some very significant results about actual infinity and infinite division were proved. In this paper 9 we will begin to see the devastating consequences of these results on the formal foundations of the physical sciences. Let us remember that there, it was demonstrated, among other, the following results:

1. ω-Ordered sets are inconsistent (Corollary 3.2).

2. The Axiom of Infinity and the actual infinity are inconsistent (Theorem 3.6 and Corollary 3.3).

3. The continuum (and then the spacetime continuum) is inconsistent (Theorem 3.8).

4. A consistent universe can only have a finite number of physical objects (Theorem 3.12).

5. Numbers with an infinite decimal expansion are inconsistent (Theorem 4.1).

6. The actual infinite division of any finite real interval is inconsistent (Theorem 4.3).

7. An interval of time can only be divided into a finite number of parts (Theorem 4.3).

8. The length of any line with two endpoints is always finite (Theorem 5.1).

9. In the Euclidean space $\mathbb{R}^3$, any closed line has a finite length (Corollary 5.1).

10. In the Euclidean space $\mathbb{R}^3$ the distance between any two of its points is always finite (Corollary 5.2).

11. In the Euclidean space $\mathbb{R}^3$ it is impossible to join any two of its points by a line of infinite length (Corollary 5.3).

12. In the spacetime continuum, the distance between any two points and the time elapsed between any two instant is always finite (Theorem 5.2).

13. There is an indivisible minimum of space (time) interval of which all space (time) intervals are an integer multiple (Theorem 5.11)

14. The laws of physics do not apply in spaces smaller than the indivisible unit of space nor in times smaller than the indivisible unit of time, both being of non-zero extension (Corollary 5.8).

15. Change is impossible in the spacetime continuum (Theorem 6.2).

Physics should prepare itself to dispense with the concept of infinity in all its models and theories of the physical world. It is rare to find a physics book (especially those related to the physics of space and time) in which the word infinity does not appear dozens of times, for example more than 240 times in [288], 170 times in [117], more than 140 times in [280], more than 70 times in [177]. And in expressions like:

... escape to infinity.

... at infinite distance.

... repelled to infinity.

... extends to infinity.

... continues to infinity.

... an infinitude of positions.

... infinitely many points.

... etc.

9.4 The spacetime continuum

As noted above, Modern Physics assumes that space and time separately are only relative, while the spacetime continuum is absolute and modeled by the set $\mathbb{R}^4$ of all real 4-tuples, (x_i, y_i, z_i, t_i) where the first three real numbers represent the spacial coordinates of a point (a spacial position) and the fourth one t_i represent an instant of time (a time position), points and instants being primitive concepts. For most physicists, neither space, nor time, nor spacetime are real physical entities but concepts necessary for the construction of models that describe reality. We will discuss this matter in paper 11.

As noted above, the spacetime continuum is made up of points and instants, which are also primitive concepts. An additional problem about these two concepts is that our intuition about them is flawed by our experience with graphic marks and the very fast events we associate respectively with points and instants. In this regard, it is useful to recall some informal definitions of point:

1. A point is that which has no part [139, p. 153].

2. A point is that which has position but not magnitude [257, p. 8].

3. Geometric element without dimensions whose position in the plane or in space is located by means of its coordinates [93].

4. A geometrical element having no dimensions; in Cartesian space, an element that can be located by a single n-tuple of coordinates [40].

5. A geometrical construct which has position but no size [56, p. 609].

So, a point would be something that, having no size (no extension), occupies a place. The problem is how a thing that has no size can occupy a place. In paper 11 it will be formally proved that, in fact, points can have neither size nor shape; and instants can have no duration. So, points and instants can only be abstract positions defined by 4-tuple of real numbers in abstract reference frames.

As just noted, the concept of instant is as primitive as the concept of point, and even more deceptive. Almost all informal definitions (much rarer than point definitions) are circular or outright wrong:

1. Very brief portion of time [93].

2. A very short time; a moment (Google Online Dictionary).

3. A very short space of time (WordReference Online Dictionary.

4. A very short period of time (Oxford Online Dictionary).

But an instant is not an interval of time, however brief the duration of that interval may be. So, as in the case of points (which have not extension), we will have to admit that instants have not duration. The same kind of formal dissatisfaction as with points:

1. Line intervals have length but a line interval only consists of a densely ordered sequence of points (between any two points of the sequence uncountable many other different points exist), none of which has length.

2. Time intervals have duration but a time interval only consists of a densely ordered sequence of instants (between any two instants of the sequence uncountable many other different instants exist), none of which has duration.

3. Any line interval has the same number of points as any other, which is also the number of points of the whole tridimensional universe.

4. Any time interval has the same number of instants as any other, which is also the number of instants of the whole history of the universe.

5. Length and duration are defined by arbitrary metrics based on one to one correspondences between points (instants) and real numbers.

On the other hand, it is important to remember that neither in space intervals nor in time intervals does immediate successiveness (adjacency) exist: no point (instant) is the immediate successor of another point (instant). Recall that in the spacetime continuum, change is inconsistent (paper 6, Theorem 6.2), just because of this lack of immediate successiveness. And not forget that physics is the science of change!

And most important of all, any continuum of points (or instants) is inconsistent (Theorem 3.8). Which makes the following question unavoidable:

> Can a theory that makes use of an inconsistent concept be consistent?

To this formal dissatisfaction must be added at least an additional physics dissatisfaction derived from the dynamics of the quantum void: modeling physical spacetime with the $\mathbb{R}^4$ continuum means that, for example, during one millisecond and on a line of one millimeter, as many virtual particles are created and annihilated as in the three-dimensional universe (with a radius of 45 billion light years) during its entire history (13.7 billion years).

As will be seen here and in the following articles of this series of articles, things are much simpler and more formally and physically satisfactory in the framework of a finite and discrete space and a finite and discrete time, both made up of indivisible and extensive units.

9.5 Problems not properly addressed

The formal basis of a science must be continuously revised, but physics does not pay attention to the formal basis of its mathematical language, nor to the lack of rigor in the use of certain key terms such as infinite, something or nothingness (See paper 2). In particular it has never questioned the consistency of the Axiom of Infinity. This is an error that could have very serious consequences. Some of them were discussed in paper 2 of this series of articles. As noted there, some words and expressions are used in physics with little formal rigor, which can lead to confusion or error, for example:

1. The word infinity is sometimes used as potential infinity and others as actual infinity, without indicating which is the case in each corresponding use. The same applies to the ellipsis "..." and to expressions as "ad infinitum," "and so on and on" etc.

2. Something and nothing as frequently used as synonymous [177, p. 7 Kindle Ed.].

3. Void and nothingness are used in a sense that is contrary to their respective original meanings.

4. How can the vacuum (which by definition has not substance) have measurable physical properties like magnetic permeability or electrical permittivity?

5. 'A fluctuation of nothingness' implies that nothingness is not nothingness but something with the capacity to fluctuate (assuming that nothingness has no capacity, not even the capacity to fluctuate).

6. 'Quantum void fluctuates continuously' means that it is not void of content but filled with something with the capacity to continuously fluctuate (assuming the void is devoid of anything that can fluctuate).

7. It is insisted ad nauseam that the velocity of a photon does not depend on the relative velocity of its emitting source, but it is almost never made clear that the velocity vector of a photon does depend on the relative velocity vector of its emitting source: the relative velocity vector of its emitting source is, or forms part of, one of its vector components (preinertia, Part 7).

8. etc.

Also relevant is the existence of problems that are systematically ignored, as if they did not exist. For example, the classical problems posed by Zeno's Dichotomy, or by change (both direct consequences of the lack of immediate successiveness, adjacency, in the spacetime continuum). Physicists act as if these problems had been solved; or as if they had no relation to physics. But they are not solved and they do have a relationship to physics. A fundamental relationship.

Recall also the questions raised in paper 7 about uniform motion that physics doesn't even ask (and don't forget that science always starts with an opportune question). Sometimes one has the impression that for some authors the strange and unusual adds value to scientific theories. I prefer simplicity. And respect for authors and theories, instead of religious veneration or contempt. We should be concerned about the almost divine position that some authors and theories occupy in the history of physics.

As noted above, another family of problems ignored by contemporary physics is that posed by the real/unreal nature of space. Indeed, for most contemporary physicists, space is not real, it is just an illusion, a fiction, like time and spacetime (see paper 11). But space is also something that expands and deform and its deformation can travel through the space itself. One could argue, as some authors do, that it is mathematical equations, and only mathematical equations, that describe deformations, expansions and vibrations of something that is not real in order to explain what happens in reality. But the question is raised again, why are these deformations necessary if there is nothing real that deforms in that way? If it is only to explain the behaviour of physical objects, that explanation would be as valid as any other explanation based on things that do not exist.

9.6 Physical space

In papers 3, 4 and 5 some important results about the actual infinity and the division into infinite parts were proved. They will be used in this section. But first, let us pay our attention to the following representative text of the hegemonic infinitism of our days [177, p. 5]:

> Likewise, our minds may not be able to easily understand infinity (although mathematics, which is a product of our minds, deals with it with remarkable success), but this does not mean that infinities do not exist. Our universe could be infinite in its spatial or temporal extent.

There is, however, one aspect of the mathematical analysis of infinity that is completely neglected: its critique. The whole mathematical theory of infinity rests on an axiom: the Axiom of Infinity. And an axiom is an axiom, not a religious creed. An axiom must be subject to criticism, especially when it is as unobvious as the Axiom of Infinity. Especially when it has behind it a conflicting history of twenty-seven centuries of discussions. In this sense, infinitist mathematics is more like a religion than a formal science:

it even has its fanatic believers who do not admit the slightest criticism and tend to respond with insult much more than with reason. I can vouch for that.

But if only one of the more than forty proofs of the inconsistency of actual infinity developed in [198] were correct, then the hypothesis of actual infinity would be inconsistent. And all the infinitist mathematics would collapse like a house of cards. The reader of this series of articles has the opportunity to analyze one such proofs: the Theorem 3.5 of the Denumerable Infinity included in paper 3. In any case, and being the existence of actual infinities established by means of an axiom, science is obliged to analyze the possibility of its inconsistency.

More than three centuries after the publication of Newton's Laws of Mechanics (Newton's Principia [240]), we still do not have a detailed formal description of motion. We will try it here, at least in the sense we are interested in. Although the description has no formal consequences on our argument about the physical nature of space (which is the main objective of this section), it is proposed as an illustrative example of the type of problems that physics hardly ever deals with. And it is also proposed as a comparative reference for the model of space, time and motion that is proposed in this and in the following articles of this series of articles.

Consider any physical object Ob with a definite shape and volume and in a definite position in the physical space PS, be PS real or unreal, absolute or relative, finite or infinite, or whatever it is if it is anything at all. Assume also Ob is uniformly moving through PS in any direction d. Let S be the spacetime continuum that models PS, and let Ot be the theoretical representation of Ob with the same shape, volume, motion and position in S as Ob in PS, but made exclusively of geometrical points densely ordered as in the continuum S. As Ot moves through S it will be occupying different points of S. So. let P be any point of Ot. At a certain instant t_1 it will be in a point P_1 of S. In order to describe the motion of Ot through S we must consider the following two exhaustive and exclusive alternatives.

1. The point P remains in P_1 a null time.

2. The point P remains in P_1 for a time τ (whatsoever it be) greater than zero.

According to the first alternative, P coincides with P_1 for zero time. But that is the same thing that happens with all the points of S

where P is not and never will be. Which is to say that the points where P is not, and never will be, are indistinguishable from the points where it is and will be. Under these conditions it is impossible to describe the motion of Ot and therefore to model the motion of Ob through PS.

According to the second alternative, at instant $t_2 = t_1 + \tau$ the point P will be in another point P_2 of S. Now then, since space is densely ordered in the spacetime continuum S, in the direction d and between P_1 and P_2 there is a non denumerable infinitude of points ($2^{\aleph_o}$ points) in which P has never been, and never will be. One could argue against this conclusion on the grounds that it is invalid because τ is not defined. But the conclusion is valid for all real numbers greater than zero, for all! Or to put it another way, there is no real number (for the value of τ) that makes this conclusion invalid. Notice that P is in P_1 for a time τ and then in P_2, without traversing the infinitely many points between P_1 and P_2 in the direction d of motion.

We must conclude, therefore, that in the spacetime continuum, either the points at which a moving object will be and not will be are indistinguishable, or its motion can only be described as a discontinuous process. A discontinuous (discrete) process in which the moving object jumps between two successive positions (P_1 and P_2 in our example) without passing through the intermediate points between those two successive positions. This proves the following:

Theorem 9.1 (of the Discrete Motion) *The continuum densely ordered spacetime cannot be used to model uniform motion.*

Let us now try to describe motion from the perspective of the Theorem 5.12 of Adjacency, according to which a qutit (qusit) can only be immediately succeeded by another qutit (qusit), so that no time elapses (space exists) between them (discrete adjacency). Remember that the Theorem 5.12 of Adjacency is a formal consequence of both the Theorem 5.3 of the Consistent Universe and the Corollary 5.8 of Discrete Threshold, according to which physical laws do not apply to lengths less than the discrete unit of length nor to intervals of time less than the discrete unit of time.

Under these conditions there will be an insurmountable maximum speed of one qusit per qutit, so that every physical object in uniform motion will only be able to advance one qusit after a certain integer number $t \geq 1$ of qutits. Where to advance a qusit means that a physical object (if it were of the size of a qusit) is in one qusit for at least one qutit after which it will be in another adjacent qusit, without ever being between the two qusits, simply because there is no qusit between two adjacent qusits. Motion would be, therefore, discontinuous (discrete) as in the second alternative above. It is also the same type of motion as that of the objects in a CALM (see next section). It is necessary, however, to formally complete this conclusion in the following terms and with the same words that were used in the article 6 of this series:

> The directional evolution of the universe demonstrates that this evolution is subject to a consistent set of rules, physical laws, (Theorem 5.3 of the Consistent Universe). So, canonical changes have to be consistent processes, and then instantaneous[1]. The problem is that we have no idea how that is possible. As a very adventurous hypothesis, it could be proposed that qusits have two modes of existence:
>
> 1. Permanence mode: the state of each qusit remains unchanged at least for one qutit. This would be the only perceptible state of qusits.
> 2. Interacting mode: all qusits update synchronically their respective states through appropriate processes lasting at least one qutit.
>
> Although, in accordance with what was said above, the problem of change will now reappear in the terms of these changes of modes. So, we would have to admit that the interactive mode is simultaneous with the permanence mode, although it remains in an imperceptible background (such as computer applications running in the background) that changes to the permanence mode at each successive qutit (or something similar).

But this kind of discrete motion is only possible if qusits and qutits are real, otherwise every physical object would have to have the ability to measure qusits and qutits by their own means in order to determine the successive positions of its motion by itself; which in turn would imply that all of them (including elementary particles) have internal clocks and rules of measurement. It seems evident that this is not the case, at least from what we know of the physical world. Consequently, we can end this section with the following important:

---

[1]As was proved in paper 6.

Subtheorem 9.1 (of Physical Space and Time) *The indivisible units of space and time are physical, and then real and absolute.*

Note that a subtheorem is not a theorem, but a statement supported by both formal and inductive elements. In this case, the formal elements are Theorem 4.3, 5.8 of the Discrete Threshold, Theorem 5.12 and Theorem 9.1. The inductive support is the enormous empirical evidence that no natural object, including elementary particles, has internal clocks and rules to measure time (qutits) and distances (qusits). Corollary

9.7 Space in CALM

In order to start building a new foundational basis for the concept of physical space, we will now recall the space in CALMs (Cellular Automata Like Models) introduced in paper 6. In these theoretical objects:

1. Space is made of indivisible and contiguous elements we call qusits (cells in the jargon of cellular automata).

2. Time is a sequence of indivisible and contiguous elements we call qutits.

3. Being contiguous means that between any two qusits (qutits) no other qusit (qutit) exists (immediate successiveness or adjacency).

4. Every region of a CALM has a finite number of qusits.

5. Every interval of time in a CALM has a finite number of qutits.

6. The state if each qusit is defined by a set of variables.

7. During at least one qutit, each qusit exhibits the same state, while in the background, the interactions between the states of the different qusits are performed, which will determine their corresponding states to be exhibited in the next qutit.

8. The changes of state are driven by the laws of the CALM.

9. Some groups qusits can be temporarily or permanently linked: they are the objects of the CALM.

Paper 10. Discrete time

Abstract.-As in the case of space, and for the same reasons, this paper 10 of the series proves the physical, and then absolute, nature of time. But time has another characteristic that has been much discussed throughout history: its reversible or irreversible nature. Making use of the Theorem 5.5 of Identicality and its corollaries, it is formally demonstrated here that time is irreversible and, therefore, that the famous arrow of time does exist in reality, an existence that is usually accepted only in probabilistic terms, and for a time that is only an illusion according to the majority of contemporary physicists. The paper also discusses on absolute motion, now within the framework of a physical space and a physical time, recalling the reasons for which absolute motion is not detectable. A summary of the properties of time from the perspective of the discrete world of CALMs is also included for comparative reference.

Keywords: physical time, absolute time, Theorem 5.5 of Identicality, irreversible time, arrow of time, preinertia, relative motion, absolute motion, time in CALMs.

10.1 Introduction

Time and instant are two primitive concepts that we have been discussing for the last twenty seven centuries. In the case of time, three aspects of its nature are discussed above all:

1. Its absolute or relative existence.
2. Its finite or infinite extension.
3. Its reversible or irreversible nature.

Contemporary science, especially physics, assumes that time is

131

relative and (potentially) reversible, i.e. irreversible in probabilistic terms with a very low probability of reversibility. Or put another way, the equations, models and theories of physics are compatible with a reversible time. And, although the observable universe has a finite age, the existence of infinite cycles of universe creation is not ruled out. Thus, as far as time is concerned, the actual infinity is still present in modern cosmology.

This paper 10 only deals with the absolute-relative and reversible-irreversible dialectic on the nature of time. Its finiteness-infiniteness has already been addressed in formal terms in papers 3-5 of this series of articles.

10.2 On the physical nature of time

The formal results demonstrated in articles 3-9 of this series will now be used to deduce some aspects of the nature of time. Let us recall some of these results:

1. The spacetime continuum is inconsistent (Theorem 3.8).

2. The infinite division of any finite distance or any finite interval of time is inconsistent (Theorem 4.3).

3. The distance (time) between two points (instants) is always finite (Theorem 5.2).

4. There is an indivisible minimum of space (qusit, quantum space unit) and of time (qutit, quantum time unit). All intervals of space (time) have an integer number of qusits (qutits) (Theorem 5.11).

5. The laws of physics do not apply in spaces smaller than the indivisible unit of space (qusit) nor in times smaller than the indivisible unit of time (qutit), both being of non-zero extension (duration) (Corollary 5.8).

6. No space exist between any two successive qusits, and no time elapses between two successive qutits (Theorem 5.12).

7. Every space interval (or time interval) is finite and can only be divided into an integer number of adjacent qusits (qutits), or into an integer number of adjacent parts, each an integer multiple of a qusit (qutit) (Theorem 5.6).

8. The indivisible units of space and time are physical, and then real and absolute (Subtheorem 9.1).

In consequence, and taking into account these results (particularly Theorem 5.6 and Subtheorem 9.1), we can affirm that time, if it is a consistent concept, can only be of a finite, discrete and real nature.

10.3 Corollary of the Composed Identicality

We will now deduce a corollary of the Theorem 5.5 of Identicality 5.5 that will be used in the next section to prove the irreversible nature of time. Recall that the Theorem 5.5 of Identicality 5.5 states:

> *All particles of the same type have the same properties and behave the same way under the same conditions.*

From this principle it immediately follows:

Corollary 10.1 (of Composed Identicality) *Theorem 5.5 of Identicality also applies to any combination of elementary particles, as atoms or molecules or even more complex systems, provided that they have the same composition and structure.*

Proof: If that were not the case, the elementary constituent particles would not be identical, which is impossible according to Theorem 5.5; or the corresponding combinations would not be identical as they are assumed to be. □

10.4 Physical time is irreversible

It is possible to interpret entropy in terms of isotropy and to formalize some inductive laws of thermodynamics [195]. From that perspective, the entropic (or isotropic) evolution of the universe is also explained in terms of the dissolution of its local anisotropies into general isotropic states. That same evolution makes possible the appearance of very local and temporary anisotropies in the form of ordered systems (such as crystallized minerals) and organized systems such as living beings. According to Kant [169, §61-84], an organized system is a teleonomic system. In the case of living beings the purpose is the same in all of them: reproduction, i.e. propagation, at least partially, of their genetic information (physical and functional information [189, 188]). Both points of view, the entropic and the isotropic, account for the same irreversible evolution of the universe.

It is important to emphasize that the formation, maintenance and evolution of ordered systems and organized systems is possible because these systems exchange matter and energy with their environment (that is why they are called *open systems*), and in such a way that the final balance of entropy (isotropy) is always positive in the sense that the joint entropy production of these open systems and their environment always increases [188, 189]. Order and organization are thus compatible with the directionality of the evolution of the universe, always in the same direction of increasing its isotropy (entropy) and its decreasing temperature.

Indeed, the universe itself and its isotropic (entropic) evolution over more than 13.7 billion years is gigantic proof of the irreversibility of time. So are the fossil record, stratigraphic series, organic evolution and the inductive laws of thermodynamics themselves. We have a gigantic empirical evidence for the irreversibility of time in our observable universe. Trillions and trillions of living beings go through the same process in the same direction: from birth to death. No exception to this directionality is known.

However, the time included in the laws of physics is reversible. It is often argued that it is a matter of probabilities in which one direction of time would be much more unlikely than the other. I think it is not much more unlikely, it is impossible. It is impossible, for instance, to reverse the processes between life and death. And living beings are part of the physical universe. I will try to confirm this inductive conviction by means of the next formal argument, which consists in proving the following two formal results:

Theorem 10.1 (of the Increasing Anisotropy) *A system can increase its anisotropy as long as the universe as a whole increases its isotropy.*

Proof: If that were not the case, either the system does not obey the Theorem 5.5, or the universe would be evolving towards states of lower entropy, which is also impossible according to the Principle 5.1 of Directional Evolution. □

Theorem 10.2 (of the Arrow of time) *Time is irreversible and its irreversibility is marked by the irreversible increase in the entropy of the universe.*

Proof: Let t_1 and t_2 be any two instants in the history of the universe. Knowing the entropy of the universe at t_1 and t_2 we also know which of the two instants is later than the other: it will be the one in which the entropy of the universe is higher (Principle 5.1). Therefore, time is directional, i.e. irreversible. □

We must conclude that time is irreversible. Contemporary physical theories do not reflect this remarkable feature of time, although everybody thinks that, in practice, time is irreversible. The problem is that contemporary physics does not care about its formal foundations, or at least does not pay enough attention to them. And that is an important deficiency, because in a poorly founded science anything can happen.

10.5 Physical time

In the article 9 on the physical space an important theorem about the trajectories of any uniform motion was proved:

> **Theorem 9.1 of the Discrete Motion**: *The continuum densely ordered spacetime cannot be used to model uniform motion.*

In this section we will demonstrate another important result, now related to any history of any physical object. For this, let Ob be any physical object in any physical state PS_1 at a given instant t_1 of the physical time PT, be PT real or unreal, absolute or relative, finite or infinite, or whatever it is if it is anything at all. With respect to the time interval that Ob remains in the state PS_1 we must consider the following two exhaustive and exclusive alternatives.

1. The physical object Ob remains in the physical state PS_1 for zero time.

2. The physical object Ob remains in the physical state PS_1 for a time τ (whatsoever it be) greater than zero.

According to the first alternative, the time during which Ob remains in the state PS_1 is indistinguishable from the time during which it remains in other states in which it is not, and never will be. Which means that the instants of the real history of Ob are indistinguishable from the instants of the non-real history of Ob.

According to the second alternative, there will be a first instant $t_2 > t_1 + \tau$ at which Ob reaches the state PS_2, otherwise, there would also be no second, third, fourth,..... instant in which O_b has already reached the PS_2 state. Now then, since time in the spacetime continuum is densely ordered, the instant $t_1 + \tau$ has not an immediate successor, and the first, second, third... instants at which Ob is already in the state PS_2 do not exist. Consequently, Ob cannot reach PS_2 at a definite first instant. One could argue against this conclusion on the grounds that it is invalid because τ is not defined. But the conclusion is valid for any real number greater than zero. Or in other words, there is no real number (for the value of τ) that makes this conclusion invalid.

We must conclude, therefore, that in the spacetime continuum, either the instants at which a physical object will be and not will be in a given state are indistinguishable, or its history can only be described as a discontinuous undefined process. A discontinuous and undefined process in which the physical object can be described in two instants at two different states, but not between all instants between those two instants. This proves the following:

Theorem 10.3 (of the Discrete History) *The continuum and densely ordered time cannot be used to model the history of physical objects.*

10.6 Absolute motion

If space and time were discrete and absolute, then motion could only be discrete and absolute. With the drawback that preinertia makes it impossible to detect absolute motion, only relative motions could be detected (see paper 7 in this series). Except, perhaps, in the particular conditions of the Santiago del Collado Experiment (SCE) whose objective is to analyze whether preinertia dominates the universality of the speed of light, or the other way around. SCE is similar to the Michelson-Morley experiment, but using a laser beam that splits into two beams, the first parallel to the direction of the dipole anisotropy caused by the Doppler effect produced by the motion of the Earth through the Cosmic Microwave Background (CMB)[1], and the second in a direction perpendicular to the first (Figure 10.1). In the first part of SCE the laser beam will be diffracted by a Fraunhofer slit, and in the second part of SCE the beam will not be diffracted. A new, much simpler and more defining version of the SCE experiment is being prepared to be conducted in early 2024.

On the other hand, and taking into account that in a discrete space-time there would be a minimum distance to travel and a minimum time to travel it, there would also exist an insurmountable maximum speed of one 1 qusit per qutit, which could be,

---

[1]According to the last data, our planet moves with respect to CMB at a velocity of 367 km/s in the direction given by the galactic coordinates $(264.4, 48.4) \pm (0.3, 0.5)$ [118], where the first coordinate is the galactic longitude (l) and the second the galactic latitude.

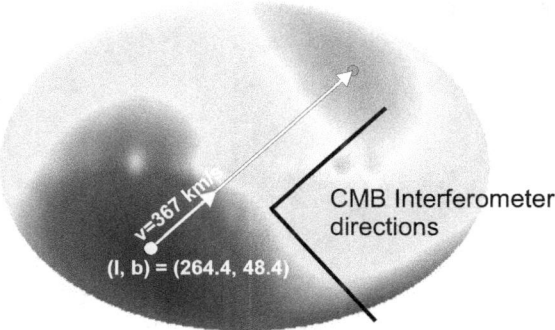

Figure 10.1 – Dipole anisotropy caused by the motion of the Earth through the Cosmic Microwave Background.

or not, the speed of light. And that, or an integer multiple of it, would also be the synchronized speed of the evolution of all qusits in the physical space, evolution directed in this case by the physical laws. The relative velocities of physical objects, the only detectable ones, would be the consequence of their different absolute velocities.

10.7 Time in CALMs

As a comparative reference, the nature of time in CALMs is recalled in this section.

1. There is an indivisible unit of time, we call also qutit.

2. Each qutit is immediately succeeded by another qutit.

3. No time elapses between two immediately successive qutits.

4. Between any two different qutits, a finite number of qutits always elapses.

5. During at least one qutit, each qusit exhibits the same state, while in the background, the interactions between the states of the different qusits are performed, which will determine their corresponding states to be exhibited in the next qutit.

6. CALMs always evolve in the same recursive way: each new state results from the application of the same laws to the immediately precedent state.

7. The evolution of CALMs is irreversible in the direction defined by its recursive sequence of states through the increasing (and never decreasing) sequence of qutits.

8. Time is irreversible in CALMs.

Paper 11. Physical versus geometrical space

Abstract.-This paper proves in formal terms a quality of all points of the continuum geometrical space that is often explicitly or implicitly assumed: points have neither size nor shape. It also highlights the contradiction of, on the one hand, denying the reality of a physical space, which continues to be (together with time and spacetime) just an illusion for most contemporary physicists, and, on the other hand, accepting that this unreal space is continuously undergoing intrinsic deformations, accelerated intrinsic expansions, and propagation of its own deformations (gravitational waves), as if something that is not real, that is just an illusion, had the actual physical ability to warp, expand, vibrate and transmit its own vibrations. If, on the contrary, the physical reality of space (together with time and spacetime) is assumed, almost insoluble problems arise related to the modeling of spacetime through a four-dimensional infinitist continuum of points without size or shape or duration. As a counterpoint, this work considers the structure and functioning of CALMs, where, apart from being free of the inconsistencies related to infinity proved in the papers 3, 4 and 5, none of these problems arise, and where it is possible to envisage a theory of gravity in which spacetime warps are unnecessary.

Keywords: sizeless points, physical points, spacetime continuum, physical space, intrinsic deformation of space, gravitational waves, expansion of space, Theorem of Formal Dependence, gravity in CALMs.

11.1 The mathematical language of physics

According to the Theorem 5.1 of the Consistent Universe proved in paper 5 of this series, the universe evolves under the control of a single set of invariant and formally consistent laws. The empirical evidence for this theorem is so enormous that until now we have not considered it necessary its formal demonstration. This series of articles demonstrates it because it is used in several arguments. Naturally, it is the formal consistency of the universe that allows its analysis and description with mathematical and computational languages.

Some theories of modern physics are actually mathematical theories, albeit scarcely concerned with the foundations of mathematics. Frequently, and with a certain arrogance, it is said and written that the universe cannot be explained with ordinary language, that it is only possible to do so with mathematical language. The problem (which physicists do not address) is what is the mathematics that explains the universe? There seems to be only one mathematics, and this is not so. Although in our days there is a mathematical current absolutely dominant over all others (infinitist mathematics), the finitist alternatives also exist.

It is true that for more than a century mathematics based on infinitist set theory has been overwhelmingly dominant. So dominant that it has ended up having a role in physics similar to that of catechisms in religions. But that does not mean that it is the ultimate and definitive mathematics. This mathematics includes a very dangerous axiom because its inconsistency would have devastating consequences on the whole mathematical edifice built since the end of the 19th century. And naturally on the physical theories most committed to this infinitist mathematics.

To put it briefly, though intentionally provocative, the Axiom of Infinity legitimizes the existence of the complete list of natural numbers, though there is no last natural number that completes the list. Or said in Aristotelian terms, the Axiom of Infinity legitimizes that the incompletable exists as completed. That statement triggered me. And thirty years and more than forty proofs later, here I am trying to convince the reader of the inconsistency of the actual infinity. So far, not very successfully, but I was warned about that.

The reader can find one such proof, the Theorem 3.5 of the Denumerable Infinity, in paper 3 of this series of articles. I chose it for its simplicity and brevity. The rest of the proofs can be found in [198]. There are many types of such proofs: based on set theory, on supertasks, on transfinite arithmetic, on geometry, etc.

Returning to the initial dictum that the universe could not be described with ordinary language but with the language of mathematics, I must express my disagreement. If physics is not expressed in the physical terms of ordinary language, then we will not have a physical description of the universe. As B. Russell would surely say, we would not know of which we are talking about [284, p. 959] [287]. But the universe is physical, not mathematical. In the same sense that our nature is physical, not mathematical. We are objects of a physical universe, not a system of mathematical equations.

In any case, prudence recommends the analysis of the foundations of infinitist mathematics. And the critique of supremacist infinitism, which so perfectly explains the universe with its extraordinary and excluding language, because in the end such extraordinary language could be inconsistent. Take a look at article 3 of this series, or at [198]. And let us not forget the arrogance of some theories of modern physics, for example special relativity, which are used as if they were new fundamental laws of logic: if something does not agree with them, that something is automatically declared false. This in the same physics that has been unable to discover something as physically evident as preinertia, surely the most basic property of all physical objects (!) The shame of physics [199, pdf].

11.2 Geometrical points and physical points

As already indicated, in some of the previous articles in this series, the set $\mathbb{R}^4$ (the set of all 4-tuples of real numbers) models the geometrical spacetime continuum that in turn models space and time in physics. The elements of that set represent points in space and instants in time, two primitive concepts (point and instant) that we make intuitive use of in most arguments about space and time in physics. As indicated in paper 9, our intuition about them is flawed by our usual way of representing them graphically.

As almost everyone knows, a simple contradiction would allow us to prove anything we wanted to prove. To the extent that one way to analyze the consistency of a theory is to prove the existence of at least one proposition that cannot be proved within that theory [194]. But, unfortunately, physicists are not very fond of questioning the foundations of their mathematical language (neither are most mathematicians, who are more religious than critical with respect to the foundations of their mathematics). Arguments such as those in this section are unheard of in physics textbooks, including the best books on physical space, such as [117].

Indeed, practically in all physics books we can find expressions such as:

1. "the space around a point"
2. "a tiny ball or point"
3. "point-like charge"
4. "mass point"
5. "mass concentrated in a point"
6. "infinitesimal point"
7. "the motion of points with mass and charge"
8. "propagation proceeds point by point"
9. "propagate through the contiguous points"
10. "creating changes at adjacent points"
11. "what happens to the field at adjacent points
12. etc.

that show the use of qualities that points do not have: extension and adjacency (immediate successiveness). The consideration that they do, leads immediately to a contradiction, as we will see here. Of course, the extension of points has never been proposed, let alone measured (otherwise we would all know it), which is almost a proof that points are assumed to have no extension. But one might think that they do, although for unknown reasons it has not yet been proposed, not even the limits that extension could have.

On the other hand, the lack of immediate successiveness (adjacency) between points in space (or instants in time) is not even considered in contemporary physics, as if they actually exist. But they do not exist: between any two points (instants) in space (time) there are always an infinity of different points (instants), none of which is contiguous (adjacent) to any other, as it happens, for example, with natural numbers: between n and $n + 1$ there is no other natural number; $n + 1$ is contiguous, adjacent to n ($\forall n \in \mathbb{N}$).

To simplify the discussion that follows, without losing an iota of formal rigor, we will consider only the points of a line r in the Euclidean $\mathbb{R}^4$ where a metric and arbitrary units of measurement have been defined, for example those of the IS system based on the multiples and submultiples of the meter. In the IS we have units of length so greater as the Yottameter (Ym) = 10^{24} m, and so small as the yoctometer (ym) = 10^{-24}m.

But we could define other submultiples of the meter inconceivably smaller than a ym using as the negative exponent any expofactorial number (see paper 4 in this series or [198] for more details), whose expression written in normal (non-exponential) text would be millions of times longer than the length of the visible universe (and they would still be numbers with a finite number of zeroes after the decimal point and before the first non-zero decimal). Since the unit chosen to express the assumed length of a point is irrelevant, we will assume that a point has length λ defined by a real number r of yoctometers (ym):

$$\lambda = r \text{ ym}, \ r > 0, \ r \in \mathbb{R} \tag{1}$$

Let us consider now three segments AB, CD and EF in a straight line L whose respective lengths are:

$$AB = 1 \text{ Ym} \tag{2}$$

$$CD = 1 \text{ m} \tag{3}$$

$$EF = 1 \text{ ym} \tag{4}$$

Obviously:

$$EF < CD < AB \tag{5}$$

We must also consider the set C of cardinals:

$$C = \{1, 2, 3, \ldots \aleph_o, 2^{\aleph_o}\} \tag{6}$$

in which multiplication is defined [48]. Cantor himself proved that any segment of a line has the same number of points as the entire line, exactly $2^{\aleph_o}$ points (Dimension Problem, see paper 4). Therefore, if we assign any length to points, for example the one defined in (1), we will have:

$$AB = 2^{\aleph_o} \text{ points} \times r\frac{\text{ym}}{\text{point}} = r \times 2^{\aleph_o} \text{ ym} = 2^{\aleph_o} \text{ ym} \tag{7}$$

$$CD = 2^{\aleph_o} \text{ points} \times r\frac{\text{ym}}{\text{point}} = r \times 2^{\aleph_o} \text{ ym} = 2^{\aleph_o} \text{ ym} \tag{8}$$

$$EF = 2^{\aleph_o} \text{ points} \times r\frac{\text{ym}}{\text{point}} = r \times 2^{\aleph_o} \text{ ym} = 2^{\aleph_o} \text{ ym} \qquad (9)$$

Consequently:
$$EF = CD = AB \qquad (10)$$

which contradicts (5). Therefore, points have not extension (size), and then they cannot have a contour, a shape. A similar argument proves that the instants of the spacetime continuum have not duration. We have then proved the following:

Theorem 11.1 (of Abstract Points) *In the spacetime continuum, points have neither size nor shape, and instants have not duration.*

But then, what qualities do have the points of the space continuum? It is often said that points only have position and that two points cannot occupy the same position. So we have to admit that positions in space are defined by objects that have no extension. Consequently, the positions would only be defined for objects, or parts of objects, that have no extension. The situation is more untenable from other physical point of view: it does not seem physically reasonable to admit that something without extension can have, for example, mass or electric charge as would be the case of mass points or charge points.

The other non-quality of points that is controversial, especially in physics, is their lack of immediate successiveness (adjacency): in the spacetime continuum no point has an immediate successor as occurs, for example, in the natural numbers, in which each number n has an immediate successor $n+1$, without there being other natural numbers between them. In the case of points, between any two of them, whatever they are, there is always the same uncountable infinite number of points ($2^{\aleph_o}$ points). If you were a point on a straight line, and you looked in either direction of that straight line, you would see no point of the line: if you were seeing a point, that point would be your impossible immediate successor.

Or to put it another way, if a point of that straight line starts to move through that line in one of its two directions, it could travel the entire corresponding semi-line without passing through any of its points: the first one to pass through would be its impossible immediate successor in that direction (this is the nuclear argument in Zeno Dichotomy [198]). It could be said that since the time of Zeno we have been warned of the discrete nature of space and time. But we have never heeded the warning. As we will see later, this situation complicates the supposed curvature and extension of the physical spacetime, when the physical spacetime is modeled by the continuum $\mathbb{R}^4$.

But the most serious problem with the continuum is its formal inconsistency (Theorem 3.8). Recall that any line of that continuum contains an infinite number of ω-ordered sequences of points, all of them inconsistent (Theorem 3.2), as was proved in paper 3. As can be seen in any of the arguments included in [198], it is not necessary to go into semiotic and abstract excesses to make a serious critique of the unacceptable inadequacies of the infinitism that has been dominant in mathematics for more than a century now. As a counterpoint, remember the simplicity of the CALMs, in which none of these problems appear.

11.3 Physical space

The dominant idea in contemporary physics is that neither space nor time are real physical objects. Answers like the next one can be found on some physics well-known FAQ websites (obviously answered by 'expert' physicists):

> Spacetime is not a fabric, it is not material. Space is just an illusion, time is just an illusion therefore spacetime is just an illusion and a good way of simplifying the concept of general relativity to the public.

This has also been the opinion of many relevant authors in the history of science and thought (particularly empiricists): G. Leibniz, D. Hume, C. Huygens, E. Mach, H. Poincaré, E. Borel, L. Wittgenstein etc. And of the vast majority of contemporary physicists. For example [302, p. 266]:

> ... space and time, like society, are in the end also empty conceptions. They have meaning only to the extent that they stand for the complexity of the relationships between the things that happen in the world.

Although, on the other hand, we can also read the contrary opinion. For instance, according to:

1. A. Einstein

   - I agree with you that the general theory of relativity is closer to the ether hypothesis than the special theory [175, p. 68].

- According to the general theory of relativity, space is endowed with physical qualities... [175, p. 98].

2. F. Wilczek:

   - Spacetime is also a form of matter [339, p. 180].
   - Spacetime has a life of its own [339, p. 180].
   - According to general relativity, spacetime is extremely rigid [339, p. 181].
   - Dark energy could be a universal density of space itself [339, p. 194].
   - What appears to our eyes as empty space is revealed to our minds as a complex medium full of spontaneous activity [338, p. 1].

3. N.A. Tambakis:

   - It seems to me that in this way we can confirm the well-known epistemological assumption that space and time are not fictions but rather modes of the dynamic existence of matter [315, p. 146].

4. M. Kaku:

   - In a sense, gravity does not exist; it is the distortion of space and time that moves the planets and stars (cited in [36, p. 63]).

The case of A. Einstein is a bit more complex. Let's remember some of his words through the years about space and time:

> 1905: The introduction of a "luminiferous ether" will prove to be superfluous inasmuch as the view here to be developed does no longer need an absolute space, at absolute rest, with physical properties [77, p. 891].

> 1913: For me it is absurd to attribute physical properties to "space" (Letter to E. Mach cited in [174, p. 135]).

> 1914: As much I am not disposed to believe in ghosts so I do not believe un the enormous thing about which you are talking and which you call space [79, p. 345].

> 1915: Thereby (through the general covariance of the field equations) space and time lose the last remnant of physical reality (Letter to M. Schlick cited in [174, p. 134]).

However, in 1916 Einstein changed his mind about the physical nature of space and the existence of the ether.

> 1916: I agree with you that the general theory of relativity is closer to the ether hypothesis than the special relativity (Letter to H. A. Lorentz cited in [174, p. 135]).

> 1919: Thus, once again "empty" space appears as endowed with physical properties, i.e. no longer as physical empty, as seemed to be the case according to special relativity. One can thus say that the ether is resurrected in the general theory of relativity, though in a more sublimated form (Morgan Manuscript, cited in [174, p. 137]).

> 1938: Our only way out seems to be to take for granted that space has the physical property of transmitting electromagnetic waves... We may still use the world ether, but only to express some physical property of space... At the moment it no longer stands for a medium built up of particles [86, pp. 159-160] [87, p. 115]

In later writings he defended that the physical notion of space is linked to the existence of rigid bodies, but he rejects the idea that space is an *a priori form of intuition* [32], as Kant defended [170]. Einstein *"always supported an objective description of physical reality, without interference of the observer"* [167, p. 128].

So, who is right? The next section proves that the attempts to solve this question leads to a serious relativistic conflict related to the real or unreal nature of (physical) space. Maybe this conflict can only be solved within a CALM perspective of space and time.

11.4 A relativistic conflict on the reality of space

Let us first consider the alternative that space is not real. The first and simplest problem would be to explain how something that can be intrinsically deformed can be unreal. One could argue that there is actually nothing that deforms, but that massive objects behave as if there is something between them that is deformed by their presence; but in reality there is nothing between them that is deformed by their presence. So, this argument is anything but a physical explanation. That is to say, we would continue ignoring the physical cause of the gravitational behavior of massive objects: the explanation given includes an *'as if'* that can be followed by any other unreal thing: as if X, as if Y, as if Z... For example, massive objects behave as if there were a God that arbitrarily determines their behavior. That doesn't look like physics. In these conditions we would have to admit again that gravitational interactions are ghostly actions at a distance, or mediated by a phantom medium.

More difficult to solve is the second problem posed by the unreal space. In the observable universe, space, and only space, expands in an accelerated way, and it did it in a super-accelerated way in the first instants of its history (inflation stage [130, 131]). Moreover, the expansion is not general, it affects certain areas of space and does not affect others, among the latter the intragalactic space. Here we are already faced with an actual contradiction: it is not possible for space, and only for space, to expand and to test in experimental terms its expansion if space is unreal. Unreal objects have no real properties. It is not logically acceptable to say that something unreal, and only something that is unreal, expands, and that this expansion is detectable in experimental terms. It seems reasonable to conclude that if X expands, and only X expands, and that expansion of X has been actually detected and measured, then X must be real. Whatever X be.

The third problem posed by the physical unreality of space is its vibration by gravitational waves (on gravitational waves there is an abundant and interesting literature: [277, 310, 309, 59, 302, 36, 268, 228, 55, 317, 44] etc.). These waves, which have already been experimentally detected, are produced by very violent gravitational interactions (for example, the collision of two black holes). Well, gravitational waves would have to be elastic vibrations of something that does not exist (the unreal space), which propagate exclusively through a medium that does not exist (the same unreal space). Moreover, they propagate through this unreal medium at the speed of light, which is a key fact, as we will see below, to explain the physical nature of space. For this explanation, we must also take into account the logical results (principles and theorems) assumed and proved in this series of articles.

Let us now consider the second alternative: space is real; it is a physical entity endowed with certain physical properties. These properties should include the following two:

1. It must permeate all physical objects, offering zero resistance to their motion (as was the case with the primitive pre-relativistic ether).

2. It can deform reversibly, and allow the transmission of gravita-

tional waves, which consist of the propagation of two mutually perpendicular series of successive dilations and contractions of space, both series in turn perpendicular to the direction of wave propagation (quadrupole radiation). This waves propagates through space at the speed of light, which requires the physical space to be a very rigid medium [339, p. 181].

Of course, both properties raise the same problem that the ancient ether posed when it was believed to be the medium through which electromagnetic waves propagated. But now it is the deformation of space itself that propagates through space itself. It is not possible then to consider that gravitational waves propagate in the vacuum (like electromagnetic waves): they are deformations of space that propagate through space itself.

It seems reasonable to prescribe that we need a new paradigm of physical space that satisfactorily explains the above drawbacks. The structure and functioning of CALMs could serve as a guideline for the construction of that new paradigm of physical space and time in which this and other fundamental problems, could be solved. The next section is an introduction to that finitist and discrete alternative.

According to the relativistic orthodoxy, the new physical space could not serve as a reference frame since it contains no perceptible elements that can be used to refer motion. The article 10 of this series addresses this issue. Here it will suffice to recall that due to the preinertia of all physical objects (including massless objects as photons), it is impossible to detect absolute motion (except, perhaps, under the conditions indicated in the mentioned article 10 of this series).

To complete the problems raised in this section on the physical reality of space, we must also consider all the formal results obtained in the preceding articles of this series, which conclude in the existence of a real, physical, space and time made of contiguous (successive) indivisible units: qusits and qutits. The steps of that argument are now summarized in the following demonstrated conclusions and proposed fundamental principles:

1. The infinity in the Axiom of Infinity can only be the actual infinity (Theorem 3.1).

2. ω-Ordered collections, and infinite collections with ω-ordered sub-collections, are inconsistent (Theorem 3.2).

3. The Axiom of Infinity is inconsistent (Theorem 3.6).

4. The actual infinity is inconsistent (Corollary 3.3).

5. The (spacetime) continuum is inconsistent (Theorem 3.8).

6. A consistent universe can only contain a finite number of physical objects (Theorem 3.12).

7. The actual infinite division of any finite real interval is inconsistent (Theorem 4.3).

8. An interval of time can only be divided into a finite number of parts (Theorem 4.3).

9. In the Euclidean space $\mathbb{R}^3$ every line with two endpoints has a finite length (Theorem 5.1).

10. In the spacetime continuum, the distance between any two points and the time elapsed between any two instant is always finite (Theorem 5.2).

11. Physical laws do not apply to lengths and interval of time respectively less than the indivisible unit of space and the indivisible unit of time (Theorem 5.8 of the Discrete Threshold).

12. The indivisible units of space and time are adjacent, so that no space exists (time elapses) between any two of those adjacent unities. (Theorem 5.12).

13. Every finite distance (or finite interval of time) can only be divided into an integer number of adjacent qusits (qutits), or into an integer number of adjacent parts, each a natural multiple of a qusit (qutit) (Theorem 5.6).

14. The indivisible units of space and time, qusits and qutits, are physical, and then absolute, real (subTheorem 9.1).

It is the latter result that establishes the real, physical nature of space and time. Note that this is a sub-theorem, a result whose proof uses formal elements and empirical inductive evidence. In this case the empirical inductive evidence is provided by natural objects, in none of which we have ever seen the existence of internal clocks to measure qutits, nor the existence of internal yardsticks to measure qusits (not to mention the very measurement procedures that all those objects would have to be continuously performing). Finally, note also the incompatibility of all these results with the spacetime continuum, and their compatibility with CALMs.

11.5 Gravity from the CALM perspective

As noted above, for most physicists space does not exist, nor does time. They are not real, they are fictitious, mere illusions. But then it turns out that the presence/absence of a massive body intrinsically deforms space, reversibly transforms it from Euclidean to non-Euclidean. And certain space deformations travel through space itself at the speed of light (gravitational waves). But this is only possible if space is a real deformable medium. It is not possible to deform what does not exist, because what does not exist has no properties, not even the property of being deformed. There are no deformations of objects that do not exist. One gets the impression that some theoretical physicists are *lost in abstraction*. The actual infinity is not a good guide.

From now on, and to avoid excessive redundancies, we will only discuss on space, though the discussion can be immediately extended to time and spacetime. As will be seen, the dialectical tension (real space versus unreal space) can only be resolved by admitting the reality of physical space, which is in accordance with all formal results listed in the above section and assumed/proved in the precedent articles of this series of articles.

We will now analyze in physical (not in mathematical) terms how a physical space modeled by the continuum $\mathbb{R}^3$ could be deformed. It has been proved above that points have no extension and no shape (Theorem 11.1). Therefore, the model $\mathbb{R}^3$ cannot be deformed by deforming its points: the points of space have no extension or shape to deform. And if the deformation of $\mathbb{R}^3$ is not possible by deforming its points, the only way to deform space would be by either by differential movements of its points, or by removing/creating points as needed.

In the first case we would have to face the problem of explain the way something that has no size can be moved. Moreover, the boundary between moving and non-moving points would also be impossible (there is no adjacency between the points of the space-time continuum). And if that were not enough, the differential motion of the points would create impossible gaps in the continuous space. In the second case, we would have to admit either the violation of the Theorem of Formal Dependence 5.6, or that the modeled universe is an open system capable of exchanging points with another unknown external reality.

The consequences of Theorem 11.1 are disastrous for the role of $\mathbb{R}^3$ as a model of a real space that can vibrate and can be continuously expanded and deformed in the Euclidean and non-

Euclidean directions. It is also important to recall at this point
the extraordinary difficulties that appear when applying the equations of gravity (both classical and relativistic) to cases of three
or more bodies. Difficulties so far insurmountable that can only
be solved with the technique of approximations and the use of
powerful computational resources. More than beautiful or ele-
gant, those equations should be qualified for what they surely
are: an approximate (and diabolically complex) way of describing
real gravitational interactions in real systems of three or more real
objects.

   Things are quite different from the perspective of CALMs. Evidently, there does not exist (yet) a CALM alternative to relativistic
gravity and its corresponding intrinsic space deformations and
vibrations. We are not even at the entrance of that alternative.
We are pointing out that there could be such an alternative. And
from that position simply indicative of a new path, we can consider
some of its peculiarities:

1. Every object in a CALM would be defined by the state of a
   certain set of qusits, being that state defined in terms of a
   determined set of variables.

2. Each object could modify the state of other CALM's qusits
   (even of all CALM's qusits), the more the closer they are to
   the object. In this sense, each object defines its own field of
   interactions (forces) with any other CALM's object.

3. Thus, what the dynamic of an object of a CALM could modify
   would not be the Euclidean/non-Euclidean geometry of space,
   but the values of the state variables of each qusit of the CALM,
   i.e. its field of interactions.

4. As discussed in Article 6, the state of each qusit is updated
   at each successive qutit thanks to the two modes of existence
   of each qusit: the permanence mode (perceptible) and the interactive mode (executed in the background, not perceptible).
   These modifications would be similar to the dynamic of a field
   of forces (including gravitational force).

5. In a CALM it would be possible to reinterpret gravity in terms
   of interactions (forces), rather than in terms of intrinsic deformations of the physical space defined by qusits.

6. Being gravity exclusively additive, if a body A of mass m_a gravitationally accelerates two other objects B and C of masses
   respectively m_b and m_c such that $m_a > m_b > m_c$, the gravitational accelerations of B and C will be the same because only

the additive gravitational pull of A counts, from which it is not possible to subtract the gravitational effects of B and C on A because gravity is only additive. This is in accord with the classical:

$$m_b a_b = G\frac{m_b m_a}{d^2} \tag{11}$$

$$m_c a_c = G\frac{m_c m_a}{d^2} \tag{12}$$

$$a_b = a_c = G\frac{m_a}{d^2} \tag{13}$$

7. Since all objects are preinertial and inertial, there would exist in all of them a fundamental mass (rest mass?) responsible for preinertia, inertia and gravitational interactions.

8. Gravitational interactions would determine the trajectories of CALM objects (including photons) through the fabric of the CALM's qusits.

Paper 7 of this series dealt with preinertia, a universal attribute of all physical objects, including photons, whereby they all inherit (in vector terms) the relative velocity of the reference frame in which they are set in motion. As one might say to a classical Greek, it's the reason we land in the same place where we jumped vertically, and not 37 km further (367 Km in modern terms). The reason for preinertia could be the rest mass, or some other universal property not yet determined. Although the most reasonable and simple thing is that it be the rest mass. If that were the case, photons would have to have some rest mass, however minuscule.

The problem with the rest mass of photons is that it makes infinity appear in the Standard Model (by breaking its gauge symmetry), but the problem is not the rest mass of photons, the problem is to make use of an infinitist mathematics language founded on an inconsistent hypothesis: the Hypothesis of the Actual Infinity subsumed in the Axiom of Infinity (see paper 3). In paper 4 a universal constant m_q was defined with the dimensions of a mass, although much smaller than Planck mass m_p:

$$m_q = \sqrt{\frac{G\hbar^3 R_\infty^4}{c^5}} = \hbar t_p R_\infty^2 =\approx 6.238883052 \times 10^{-64} Kg \tag{14}$$

where R_∞ is Rydberg universal constant. Being defined in terms of universal constants, m_q is also a universal constant, be it or not the rest mass of the photon. This mass m_q can also be written as a fraction of m_p:

$$m_q = \frac{G m_e^2 e^8}{8^4 \pi^6 \epsilon_o^4 \hbar^5 c^5} \, m_p = 3.146 \times 10^{-57} m_p \qquad (15)$$

where m_e is the rest mass of an electron, e the unit of electrical charge, and ϵ_o the electric permittivity. Be it or not the rest mass of a photon, m_q is in the order of magnitude of other estimations, most of which range from $< 10^{-51}$g to $< 10^{-64}$g [322, 299, 326, 110].

11.6 Expanding geometrical space and physical space

Apart from the initial inflationary stage, the physical space of the universe, and only it, is expanding since its formation, and it is expanding faster and faster. Furthermore, there are areas of space that are expanding and areas that are not expanding, such as intra-galactic space. Here we encounter the same problem of space deformation discussed above. How can something that has no real existence, something that is only a fiction, an illusion, be expanding for more than 13.7 billion years? As in the case of geometric deformation, we will have to admit that if space expands it is because it is something with the physical capacity to expand. And only real objects have physical capabilities. Therefore, space must be real. Even if it is modeled by the continuum of real numbers.

If physical space is modeled by the set $\mathbb{R}^3$, then the physical version of the points of $\mathbb{R}^3$ must be real physical elements that have neither size nor shape. Consequently, the expansion of the universe cannot be caused by expanding its points, because these would cease to be points. Nor can space be expanded by creating gaps between its points, in this case the continuum of points would cease to be a continuum to become a discontinuum with gaps that over time would grow in number and/or size.

The only solution is that new physical points appear (whatever these physical points modeled by the geometric points of $\mathbb{R}^3$ are). And here problems also appear because that continuous creation of physical points would imply one of the following two alternatives:

1. The universe violates the Theorem of Formal Dependence 5.6 according to which no formal object is self-defining, self-proving,

or self-causing.

2. The universe is not an isolated system, and there would have to be another unobserved reality from which comes the new space that makes expansion possible.

By contrast, from the discrete and finitist perspective of a CALM, it does not seem necessary any expansion of space, it would be enough to analyze the possibilities of motion of CALM's objects through the space defined, once and for all, by the CALM's fabric of qusits.

The physical reality of space deduced from its ability to deform, vibrate and expand raises the question of absolute motion, which is anathema to modern physics. But if physical space is real, then why shouldn't it be possible to move THROUGH it? Paper 10 deals with this issue.

11.7 Fields and CALMs

One of the most fruitful and relevant concepts in the history of physics is the concept of field. Although the basic idea of a field can already be found in Leibniz, it was explicitly introduced by Faraday (an experimental physicist with little mathematical training). Shortly after, Maxwell expressed the electromagnetic field in mathematical terms with his famous equations. Since then, the use of the concept of field has been generalized in almost all areas of physics (theoretical and experimental) with remarkable success. The Oxford Physics Dictionary and The Oxford Philosophy Dictionary give the following definitions:

1. **field**: A region in which a body experiences a force as a result of the presence of some other body or bodies. A field is thus a method of representing the way in which bodies are able to influence each other [64, p. 184].

2. **field**: A central concept of physical theory. A field is defined by the distribution of a physical quantity, such as temperature, mass density, or potential energy, at different points in space [34, p. 134]

There are two basic ways of looking at the concept of field:

1. A physical medium from whose variations result the interactions of the objects contained in that physical medium.

2. A way of describing the way in which different physical objects interact with each other, without there being an actual physical medium from whose variations these interactions might result.

Faraday was in favor of the first alternative. For him, the similarity between different fields was a proof of the physical reality of the corresponding media. For instance [96, 3284][97, p. 20]:

> All these effects and expedients accord with the view that the space or medium external to the magnet is as important to its existence as the body of the magnet itself.

Faraday's view invites to consider the possibility of a reinterpretation of physical fields from the point of view of the structure and functioning of CALMs. An interesting possibility would be the development of a quantum field theory within a CALM.

Paper 12. On space deformations

Abstract.-This article in the series examines certain problems related to the contraction-expansion of space that have not yet been considered by contemporary physics. It examines how these contractions and expansions would have to be carried out given the formal properties of the set $\mathbb{R}^3$ of 3-tuples of real numbers used as a model for the space continuum, in particular the dense order of the set of real numbers, from which the impossibility of immediate successiveness (adjacency) between its elements is derived: between any two real numbers there is always an infinity of other different real numbers. This simple and well-known numerical fact imposes certain restrictions on how space intervals can be expanded and deformed, which poses new difficulties for physical theories that make use of such expansions and deformations of space.

Keywords: space expansion, space deformation, length contraction, set densely ordered, immediate successiveness, adjacency, real numbers, real intervals, Theorem of Formal Dependence, self-creation, self-destruction.

12.1 Introduction

The deficient use of formal language, even ordinary language, in contemporary physical theories has already been discussed in previous articles in this series, especially in the first two of them. Indeed, the inappropriate (because inaccurate) use of expressions such as adjacent points, contiguous points, point-to-point, etc. are not uncommon in the primary literature of physics, not to mention its secondary literature. The inappropriate use of certain terms and expressions has serious consequences, providing

157

authors with a distorted and erroneous view of some foundational aspects of physical sciences. In T. Maudlin words [217, p. xiv]:

> Unfortunately, physics has become infected with very low standards of clarity and precision on foundational questions, and physicists have become accustomed (and even encouraged) ti just "shut up and calculate," to consciously refrain from asking for a clear understanding of the ontological import of their theories.

The same is true of certain mathematical and geometrical concepts that are essential to the construction of physical theories. These conceptual errors prevent them from seeing the real problems posed by the infinitist objects they use in their mathematical modeling of the physical world. And the problems are so serious that they call into question the physical theories themselves. All this apart from the formal inconsistency of the Actual Infinity Hypothesis.

The deficient use of language, and even of logic, in physical theories has already been denounced by other authors, and in a very solvent way. This is, the case, for example, of the article 'A Bang into Nowhere' by C. Antonopoulos, from which I include the following quotes on the expansion and deformation of space, somewhat related to the content of this article [42]:

> The growing (or expansion) of the balloon *presupposes* space. But the growing of Space cannot similarly presuppose space because, presumably, Space itself is being created *by* such growing (or expansion) and is presently just as large as it has grown, and not more [42, p. 49].

> If Space is curved, and therefore if Space has a shape, then *Space* occupies some parts of Space but not some *other* parts of Space. Or, somewhat differently, Space can be found only in *some* points of Space but not in *all* the points of Space [42, p. 55].

Over the last century we have become accustomed to strangeness in scientific theories (strangeness that is often confused with difficulty), to the point of evaluating the strangeness content of scientific theories as positive, as if this strangeness added some kind of positive and attractive value to them. Some theories presume to be strange, and therefore difficult to understand. I wonder what will happen when it is discovered that this strangeness actually results from the inconsistency of some of the foundational hypotheses of such theories, as is the case of the Actual Infinity Hypothesis that underlies the mathematical language of contemporary physics, a language that physics never questions. The lack of formal rigor in the case of the (primary and secondary) literature on the expansion and deformation of space can reach levels that, according to Antonopoulos, are already very close to the ridiculous.

12.2 Points have neither extension nor shape

As noted in Paper 4 the firsts Pythagorean believed in the existence of indivisible geometrical points with an extension δ greater than zero [213, pp. 11-16]. But the Pythagoreans themselves later discovered the existence of incommensurable lengths, and with them the extension of points disappeared forever. In effect, since then we suppose that points have no extension (and instants have no duration). And by means of a little transfinite arithmetic, it is possible to prove in formal terms that this is the case. Recall Theorem 11.1 proved in Article 11:

> **Theorem 11.1.**-*In the spacetime continuum, the points of space have neither size nor shape, and the instants of time have not duration.*

Unfortunately, Pythagoreans thinkers did not have the occurrence of integer division and the subsequent possibility of a discrete arithmetic. On the contrary, the era of the continuous space and time was inaugurated. Confirmed, moreover, by our sensory perception of the physical world as a continuous world. Think, for example, of the continuous perception of motion. The invention of cinematography, more than twenty-five centuries later, came too late and our science is completely based on the continuum of the real numbers that models spacetime. The problem is that that spacetime continuum is inconsistent as was proved in Theorem 3.8 of the Inconsistent Continuum. And this will change all, because no scientific theory can make use of an inconsistency as one of its fundamentals.

12.3 Experiments and theories

Educated as a naturalist, I am a strong advocate of experimental science, although experiments are also subject to error, almost always related to precision. It is experimental results (including all kinds of observations) that should guide the development of natural sciences. In this regard, it seems appropriate to recall the following points:

1. Experimental results may be compatible with more than one theory. For example, the first experimental confirmation of time dilation resulted from the Ives-Stilwell experiment, which was compatible with Einstein's theory of special relativity, and with H. E. Ives' own theory (absolute space and time) [158, 161, 162, 324, 323]. This last detail is one that is almost always forgotten.

2. Experimental results are also compatible with the apparent, not real, nature of certain physical phenomena: we can experimentally measure the refractive deformation of a rod partially immersed in water and confirm Snell's Law over and over again, but the rod is only apparently deformed, not really deformed.

3. The relativistic deformations (special relativity) of material objects, or of some mechanisms (e.g. clocks), do not depend on the nature of the deformed objects (or on the type of mechanisms in the case of mechanisms): they only depend on the relative velocity at which they are observed; i.e. they do not depend at all on the deformed object itself but on the way they are observed from outside the object. But this is also the case for refractive deformations. And as with refractive deformations, relativistic deformations could also be only apparent.

4. In some cases, such as the relativistic space contraction, or the relativistic time dilation, the experimental confirmations must be double and symmetric, but almost never are: if from a reference frame A it is observed the contraction of a ruler in another reference frame B, then from the reference frame B the same contraction of the same type of ruler must be observed in the reference frame A. And the same applies to inertial time dilation and inertial phase difference in synchronization.

5. Experimental confirmation of a theory can also occur if it results from interpreting an essentially discrete physical world (including space and time) in terms of continuum-based mathematics: the relativistic Lorentz factor coincides with the conversion factor between the discrete and the continuum versions of Pythagoras Theorem.

6. In any case, we should avoid that some scientific theories, such as the theory of special relativity (SR), end up usurping the role of the fundamental laws of logic, as is already happening: if this or that statement is not compatible with SR, then that statement is false. This is equivalent to equating SR with the fundamental laws of logic.

All of which suggests a humble and not arrogant attitude in the defense of scientific theories, however experimentally proven they may be.

12.4 Expanding and contracting the space continuum

With regard to the deformations of space, and in spite of the arrogance with which contemporary orthodoxy is defended, the reader should remember that things are far from being resolved: there is a deep division among its proponents (see previous article on the physical/geometrical nature of space):

1. Those who believe that space is unreal, perhaps the majority of authors: space is just an illusion; space is a fiction: space is an empty conception; etc.

2. And those who defend its reality (like the second Einstein from 1916) [174, 175]): space is endowed with physical qualities [175, p. 68]; spacetime is also a form of matter [339, p. 180], etc.

And, obviously, the division is very significant from the point of view of the space deformations: can a non-real object be deformed and the deformation be empirically measured?

Contemporary physical theories are essentially mathematical theories, modelizations in the form of equations of different types (linear, nonlinear, differential, in partial derivatives etc.) developed always with the same language of infinitist mathematics, the mathematics based on the Hypothesis of the Actual Infinite subsumed in theAxiom of Infinity. But one thing is the model and another the physical world. It seems appropriate to recall here again the following words of P. Dirac written more than 60 years ago [71, p. viii] (cited in [41, p. 11]:

> Mathematics is only a tool and one should learn to hold the physical ideas in one's mind without reference to the mathematical form.

But the most relevant aspect of the expansion (contraction) of the space modeled by the set $\mathbb{R}^3$ of all 3-tuples (x, y, z) of real numbers is the existence of certain restrictions imposed by the dense order of the real numbers, and consequently by the impossibility of adjacency (immediate successiveness) between the elements of that set, as well as by the null extension of the points of the continuum R^3 (Theorem 11.1). Among those restrictions, we have to consider the following:

1. Points cannot be deformed because they have no shape to deform. Therefore a space of points cannot be deformed by deforming its points.

2. Points have no extension:

    a) Therefore a space of points cannot contract by the contraction of its points: points have no extension to contract.

    b) Neither can the space be stretched by stretching its points: those points would cease to be inextense, and therefore would cease to be points.

3. A space of points cannot be deformed by sliding its points: they would leave gaps of indeterminable extension and space would no longer be a continuum of points.

4. Furthermore, collisions between points would occur, which is impossible because the collided points would have to be contiguous, which is not possible because contiguous points are impossible in a densely ordered set of points.

5. The deformation of a space of points cannot occur either by destruction of points or by creation of new points:

    a) In the first case, gaps of indeterminable extension ($0 \times 2^{\aleph_o}$ is a mathematical indeterminacy) would be created and the continuum would cease to be a continuum.

    b) In the second case, there would have to be gaps of indeterminable extension for the new points, which is impossible in a continuum.

6. A deformed space would be indistinguishable from an undeformed space: in any direction of the space, deformed or not, there would always be the same number of inextensive, formless, non-deformable and non-adjacent points: always $2^{\aleph_o}$ of such points.

Consequently, it seems impossible to deform a space composed of points without size, without shape, without contiguity, and of which the same number would exist in any region of the universe, whatever its size, from a Planck volume to the entire three-dimensional universe.

12.5 The relativistic contraction of space

In the year 1889 G. F. FitzGerald [133], and in 1892 H. A. Lorentz [226], proposed independently a real length contraction of moving objects in the direction of absolute motion through the ether in order to explain the negative results of the Michelson-Morley experiment (in the case of FitzGerald it would be an expansion in the orthogonal direction [201]). According to both authors, the contraction was caused by changes in the intermolecular forces of the moving objects as a consequence of their interactions with the ether, even though there was no empirical evidence for these changes.

Some years later, in 1905, A. Einstein writes on the different lengths of a rigid rod when measured at rest and in relative uniform motion [85, p. 95]. This relativistic contraction is immediately deduced from the Lorentz Transformation [197, pp. 70-71]. It is the well-known FitzGerald-Lorentz contraction, according to which a straight line AB in its proper reference frame is observed contracted by a factor f when its length is measured in relative motion:

$$f = \gamma^{-1} \cos \alpha_o = \sqrt{1 - k^2} \, \cos \alpha_o \tag{1}$$

where γ is the Lorentz's factor; $v = kc$, $0 < k < 1$, is the relative velocity; and α_o the angle that the straight line AB makes in its proper reference frame with the direction of the relative motion. And that is all.

As can be seen, nothing is said about the way in which the line is contracted; the restrictions and possibilities outlined in the previous section are not mentioned. If we change the line AB for a physical object, for example a metal ruler R, maintaining the conditions of the relative velocity motion, we would see the ruler R with all its marks and numbers contracted in the direction of v by the same above factor f.

The problem with the rod R is that it can be observed with a multitude of different relative velocities, and therefore the measurement of its length will be different for each of these measurements. This opens the question about the real or apparent nature of the FitzGerald-Lorentz contraction, because it is difficult to assume that the same physical object can simultaneously have so many different lengths. Today there are supporters of both alternatives. Those who defend the reality of contraction would have to explain:

1. How is it possible that all physical objects, whatever their composition and internal structure, contract in exactly the same way exclusively dependent of the relative velocity at which they are observed?

2. What is the physical cause of the contraction? the contraction of atoms? the contraction of interatomic distances?

3. How can the same physical object have different sizes at the same time?

4. Can there be as many simultaneous superimposed physical realities as there are different velocities at which their objects and events can be simultaneously observed?

5. FitzGerald-Lorentz contraction is incompatible with some well established physical laws and empirical data, for instance [197, pp. 64-355]:

   - It is impossible for an elastic cord free of forces to be enlarged only in some of its parts.
   - The laws of mechanics governing the deformation of rigid materials as glass cannot be violated.
   - It is impossible to deform continuously a metal object without any force act upon it.
   - It is impossible to make disappear the rotation of a physical object.
   - It is impossible for the hydrostatic pressure not to be the same in all directions.
   - All optical isotropic media behave as anisotropic when observed in relative motion.
   - The speed of a photon through a transparent medium should be the same in all reference frames, but it is not.
   - Optical isotropy cannot be observed in relative motion.
   - Optical isotropic material only exist at rest.
   - The relativistic bipolar anisotropy does not exist in real media at rest.

- The laws driving the kinetics of oscillating reaction cannot be violated.
- Two identical clocks identically accelerated should not behave in different ways while they are accelerated.
- Extreme anisotropy is unknown in optical crystallography.
- It is impossible that a simple reflection of a photon on a mirror changes the speed of the reflected (re-emitted) photon by hundred of thousands km/s.
- The second law of the reflection of light can be violated in certain relativistic circumstances.
- Snell Law is also violated in certain relativistic conditions.
- The total internal reflection of light can also occurs under anomalous relativistic circumstances.
- It is impossible for a rigid body to enlarge indefinitely.
- It is impossible that an explosion occurs and does not occur.
- It is impossible that a standard ideal pendulum swings faster in one direction than in the other.
- It is impossible to accelerate a physical object without a force acts on it.
- It is impossible to accelerate photons if they always moves with the same universal speed.
- The impossibility to describe motion with respect to external references does not imply that motion does not exist. It necessarily existed without observers who could describe it for the first few million years of the universe's history.
- The universality of the laws of logic is incompatible with the lack of relativistic simultaneity.
- etc. (see [197])

Those who defend the apparent nature of the FitzGerald-Lorentz contraction have also a some questions to answer:

1. If time dilation and phase difference in synchronization (local simultaneity) are also deduced from the same Lorentz Transformation as the FitzGerald-Lorentz contraction, are they also apparent?

2. If the three relativistic effects can be experimentally confirmed and deduced from the same Lorentz Transformation, why the FitzGerald-Lorentz contraction would be apparent and the other

two real effects? Where does special relativity (SR) establish that some inertial spacetime deformations are real and others are apparent?

3. If the three effects are only apparent, would not SR be explaining a reality that is observed to be deformed because of relative motion but is not actually deformed?

4. If that were the case, would not SR be a theory of apparent deformations of little interpretive value of the real physical world?

12.6 The expansion of intergalactic space

As is well known, the experimental verification of the expansion of the universe predates the Big-Bang theory, which was precisely inspired by that expansion. What we also know at least since 1998 is that this expansion is accelerated. The theory that tries to explain the facts concludes that it is space itself that expands, as in the original times of the great cosmic inflation, but in a more moderate way: at a speed of about 70 Km/s per megaparsel. Moreover, not all space expands, the space inside the galaxies does not: the objects of a galaxy maintain their relative distances. Naturally, the general theory of relativity is an essential part of the theoretical framework that explains the expansion of space, which naturally excludes any possibility of absolute motion, and therefore of absolute space and absolute time.

But as we have already seen in this series of articles, on the very existence of space as a physical entity in its own right, there is a deep division of opinions. For some authors (the majority) there is no such physical entity:

> It is only an illusion. A good way of simplifying the concept of general relativity to the public (see paper 11).

For others, space exists as a physical entity:

> A form of matter [339, p. 180]. A complex medium full of spontaneous activity [338, p. 1]. Modes of the dynamic existence of matter [315, p. 146], etc. (see paper 11).

Consequently, and on the issue of the expansion of space, we would have to highlight the following:

1. For the supporters of the non-existence of physical space:

    1. The universe evolves *as if* there were something (which in reality does not exist) that expands according to certain equations. This attitude is compatible with the consideration of any other object that does not exist but has the appropriate properties to explain whatever one wishes to explain, and to assume that, indeed, that is the explanation one is looking for. These objects are rejected by formal logic as objects defined on purpose (*ad hoc*) to achieve the desired end.

    2. The problem with imaginary objects is that they do not physically exist, and trying to explain the evolution of what does physically exist with objects that do not physically exist is not physically acceptable; or at least it is not a complete physical explanation.

    3. If we assume that physical objects in the universe interact only with other physical objects in the universe, it is not possible to explain the evolution of the universe by the interaction of real physical objects with imaginary non-real objects. It seems reasonable to require that reality must be explained only by the properties of real objects.

    4. Even though it is imaginary, this unreal space would have the ability to create itself, which goes against Theorem 5.6 of the Formal Dependence:

        Theorem 5.6 (of Formal Dependence).-*No concept defines itself; no statement proves itself; no physical object is the cause of itself; and no cause is the cause of itself.*

2. For the supporters of the existence of physical space who assume the physical space is properly modeled by the continuum R^3:

    1. The points of the continuum R^3 must represent something like the physical points of physical space. A physical point would have to be a physical object (not a quality of a physical object). But how can something that has no extension be a real physical object?

    2. Any region of physical space has the same number ($2^{\aleph_0}$) of inextensive physical points before and after any expansion. Under these conditions, how is the expansion of physical space possible if neither the number of its points nor the size of its points increases?

3. The expansion of physical space requires its capacity for self-creation, which goes against Theorem 5.6 of the Formal Dependence.

12.7 The gravitational deformation of space

Although there are alternatives, the hegemonic theory of gravity in contemporary physics is the one based on the general theory of relativity. Gravity would not be a force but the consequence of a geometrical deformation of space: physical objects, including photons, would move along geodesics that can be straight lines or lines curved by the nearby presence of objects with the appropriate mass.

To those who defend the unreal nature of physical space it makes sense to pose a couple of questions:

1. How is it possible to deform something that does not exist?

2. How is it possible for physical objects to move along non-existent geodesics? What quality of the object determines the next (non-existent) point to which the object is to move?

Those who defend the real nature of physical space can be asked the same questions they were asked in the case of the expansion of space, and they will have to draw the same conclusions: their theory is incomplete and will remain incomplete because of the impossibility to describe how these deformations occur in fact. They can only quantify them.

In any case, it should be noted that with respect to gravitational deformations other much simpler theories could be developed based on preinertia, and in which it is not necessary to deform any space, be it a real space or an imaginary space. Since preinertia is a universal property of all physical objects (including photons), there must exist in all of them something common that is the cause of that preinertia. The simplest explanation could be that that something is mass (an undefinable primitive concept), the same that originates the inertial mass and gravitational mass of all material objects. Thus, although much remains to be discussed on mass, the simplest explanation of the above three facts would be that all material objects have a property capable of modifying the properties of space (gravitational fields), of offering a certain resistance to their own changes of motion (inertial mass), and of inheriting (in vector terms) the motion of the body from which it starts its own independent motion. This latter property manifests itself even in supposedly massless objects such as photons. There is a possibility that photons have a rest mass of the order of 10^{-64} Kg, which could be called quantum mass m_q [197, p. 235]:

$$m_q = \sqrt{\frac{G\hbar^3 R_\infty^4}{c^5}} = \hbar\, t_p R_\infty^2 = 6.238883052 \times 10^{-64} Kg \qquad (2)$$

where t_p is the Planck time and R_∞ is the universal Rydberg constant, which is specific to each chemical element and varies slightly with its mass.

Paper 13. On time deformations

Abstract.-The continuum $\mathbb{R}^+$ modeling physical time, real or not, poses the same problems in the case of time as in the case of physical space discussed in the article 12 of this series. In the case of time, moreover, the continuous model of time makes impossible the solution of one of the oldest physical problems that remains unsolved: the problem of change (introduced in paper 1 and discussed in paper 6 of this series). The main objective of this article 13 is the discussion of the real or apparent nature of relativistic time deformations. Here, it is proved that, despite their experimental confirmations, relativistic time dilation and local simultaneity (phase difference in synchronicity) could be apparent, not real, in the same way that all refractive deformations are apparent, no matter how many experiments confirm Snell's Law. This paper also demonstrates the impossibility of describing relativistic time dilation in terms of the successive instants that supposedly define the passage of time. The discrete and real nature of time, already demonstrated in previous articles of this series, not only indicates a way to solve the problem of change, but could also explain all relativistic deformations of space and time.

Keywords: time dilation, difference in phase synchronization, set densely ordered, immediate successiveness, adjacency, real numbers, real intervals, twin robots paradox, impossible pendulums.

13.1 Introduction

We will never know what would have happened if the pre-Socratics, who initially considered that the points of space had a non-zero extension before discovering the incommensurability of the side of a square with its diagonal, had also discovered integer division and discrete arithmetic. Or that the discrete view of the physical world (including space, time and motion) introduced in the tenth century by the Islamic school of thought Kalam [165, p. 62-68] would have been accepted and developed by the modern science that emerged in the sixteenth century (see paper 4). But neither possibility occurred. What did occur was the appearance of the irrational numbers, and eventually the set $\mathbb{R}$ of the real numbers and the actual infinite, numerable and non-numerable. In my opinion, with Cantor, science entered a phase of infinitist bewitchment in which it still remains, and from which I believe it will be very difficult to emerge. Like the Pied Piper of Hamelin, the actual infinite has been guiding the steps of mathematics and physics for the last century and a half.

The article 3 in this series provided the reader with a very simple proof of the inconsistency of the actual infinity, and hence of the spacetime continuum (Theorem 3.8). The inconsistency of the actual infinity should change everything, although it will do so by making many difficulties. In the meantime, this article examines some of the consequences of using the set $\mathbb{R}$ of real numbers as a guide and model. Some of the aspects that were discussed in the previous article concerning space deformations are discussed here concerning relativistic time deformations. The discussion leads to results that are unacceptable from both the logical and the physical points of view. The solution could be to assume these deformations are not real but apparent. Or, alternatively, to assume that space and time are both discrete, discontinuous.

13.2 Instants have neither duration nor contiguity

Whether real or not, physical time in contemporary physics is represented (modeled) by the continuum $\mathbb{R}^+$ of positive real numbers. And it is within that model that the Theorem 11.1 was proved. According to this theorem the instants of time have no duration, nor do the points of space have extension. Neither do they have immediate successiveness (adjacency, contiguity): between any two instants there is always an infinite number of different instants, successively increasing, but without any of them being the immediate successor of another; as it happens with the natural numbers, where any of them, n, does have an immediate successor $n + 1$ (Peano's Axiom of the Successor [252, p. 1]), and so that between n and $n + 1$ there are no other natural numbers.

That topological feature of $\mathbb{R}^+$ (the lack of points adjacency) raises numerous problems related to the way in which space and time deformations could be carried out. In the previous article it was shown that such deformations cannot occur:

1. By deforming the points (instants)

2. By changing the size of points (instants).

3. By adding or removing individual points (instants).

They are, therefore, formally impossible. In the case of relativistic time dilation and relativistic local simultaneity we know only the global factors:

$$\gamma = (1 - k^2)^{-1/2} \qquad \text{(dilation of time)} \tag{1}$$

$$\frac{\gamma L_o k}{c} \qquad \text{(phase difference in synchronization)} \tag{2}$$

that say nothing about how the deformations occur ($v = kc$ is the relative velocity ($0 < k < 1$), and L_o is the proper separation between two events in the direction of the relative velocity v).

13.3 The model R^+ of time and the problem of change

The problem of change was introduced in the paper 1 and discussed in the paper 6 of this series, where the Theorem 6.2 of Change was proved:

Change is impossible in the spacetime continuum.

It was also shown in the aforementioned article 6 that immediate successiveness (adjacency) is a necessary condition to consistently explain any canonical change of any nature. Obviously, adjacency is possible in discrete time, with minimal indivisible and adjacent units of time (qutits), so that the problem of change could find a solution in the framework of a discrete space and time as proposed in this series of articles. And remember that, however forgotten the problem of change may be, until it is solved we will not have a sufficient explanation of the physical world, precisely because the physical world is in continuous change.

(The following three sections are taken from [197])

13.4 The Ives-Stiwell experiment

The first suggestion that the ticking of clocks could change with movement was published in 1887 [330]. And perhaps the first suggestion of how this alteration could be verified experimentally came from a debate between A. Einstein and W. Ritz, but at that time the experiment was considered unfeasible [163]. Some twenty years later, in 1938, the experiment was performed by H. E. Ives and G. R. Stilwell [163, 157, 156, 159, 160], a short modern review can be found in [98].

Ives-Stilwell used a Dempster tube of canal rays and the Transversal Doppler shift of the radiation emitted by the moving particles. The spectrographic analysis produced results consistent with Larmor - Lorentz predictions. Or in other words, compatible with Ives's theory of absolute space and time and with Einstein's theory of relativity. As is well known, after the Ilves-Stilwell experiment, many different experimental verifications of (the relativistic) dilation of time have been carried out. But here we are interested in the first one for the following reasons:

1. It so happens that, invariably, whenever the Ives-Stilwell experiment is cited, it is done to indicate that it was the first experimental verification of Einstein's theory. But it is (almost) never noticed that it was also an experimental test of Ives's theory of absolute space and time. Einstein knew the Ives-Stilwell experiment and its results, although he never cited them (just as he never cited the Sagnac effect or the Michelson-Gales experiment) [323, p. 44, 84-85]. By the way, with that same theory, Ives was also able to explain the anomaly of Mercury's orbit, previously explained by Einstein in 1916, but in the case of Ives with different methods of classical mechanics that do not assume the curvature of space-time.

2. We are then in front of experimental results that are compatible with two very different theories (and surely with others): Einstein's theory of relativity and Ives' theory of absolute space and time. A significant detail forgotten by the most fervent believers of Einsteinian relativism.

3. As with most time dilation experiments, the Ives-Stilwell experiment takes place within the strong gravitational field of the Earth. This detail should be taken into account when it is stated that these experiments demonstrate the inertial (exclusively due to rectilinear and uniform motion) dilation of time. In any case we have to remember again that a verification of

the inertial dilation of time requires two things: to verify from an inertial reference frame A that the clocks of another inertial reference frame B, both in non-zero relative motion, run slower than A clocks. And at the same time, check from the inertial frame of reference B that the clocks of A go slower than those of B. If this is not the case, the proof is incomplete.

This double requirement is at the base of Herbert Dingle's critique (see Chapter 52) which, in my opinion, has not been satisfactorily answered. In short: if there is only a unique objective reality, then clock A ticks and does not tick slower than clock B; and clock B ticks and does not tick slower than clock A. And if there is no single objective reality, there would have to be as many simultaneous and superimposed realities as there are relative ways of observing the clocks A and B.

On the other hand, it is appropriate to emphasize again that the relativistic dilation of time has to be universal: the same for all kinds of clocks. Mechanical, electrical, electromagnetic, electronic, atomic, biological clocks... all of them would undergo the same increase in the duration of the periodic events by which they measure time. And this raises an unavoidable question (although not very frequent in relativistic catechisms) related to that universality: how is it possible that so many different objects (in their chemical composition, in their internal structure, in their periodic mechanisms) all suffer the same increase in the duration of the periodic events by which they measure time when they are observed in relative uniform motion? And let us not forget that for most physicists time is only an illusion, a fiction to express certain relationships between physical objects.

As a comparative reference, let us consider any number of rigid rods of the same dimensions, immersed partially in water under the same conditions: all of them will show the same deformation regardless of their chemical composition and internal structure (glass, plastic, copper, aluminum, ceramic, wood etc.). That under the same conditions exactly the same deformation occurs in such a great variety of materials suggests that, in fact, these observed deformations are only apparent.

And these time dilations being exclusively related to the relative velocity at which the clocks are observed, we would have to admit, moreover, that if these time dilations were real, all clocks of all imaginable types would have to function simultaneously in as many different ways as there are different relative velocities at which they can be observed. Or, alternatively, they should start running at a certain rate when they begin to be observed with a new relative velocity; and stop running at a certain rate when they cease to be observed with a certain relative velocity (uniform velocities in all cases). In addition, and in some way, the respective clocks must receive the information that they begin (or end) to be observed at a certain relative velocity. Which is still a rather twisted conclusion and alien to the simplicity of all known natural laws.

Taking into account the enormous diversity of all existing and imaginable clocks, the answer to the question on how they can all modify their respective ticks in exactly the same way exclusively dependent on the uniform relative velocity at which they are observed may have to be sought at a more basic level than the level of physical laws, or simply from the perspective of appearances (as in the case of refractive deformations). And from the purely descriptive point of view: how can a densely ordered succession of instants be dilated if its instants cannot be dilated and their number cannot increase either: it will always be $2^{\aleph_o}$, both in one second and in the whole history of the universe?

13.5 Relativistic dilation of time

The following argument is reminiscent of the famous twin paradox, although it contains new elements that allow different conclusions to be drawn. Indeed, consider three *identical* robotic observers, the robot A, the robot B and the robot C, in their proper frame RF_o, and assume they are programmed to carry out the following tasks:

a) At the instant t_{oo} of RF_o, the robots A and B accelerate in exactly the same conditions until each of them reach the same uniform speed u with respect to RF_o. The only difference is that A and B are accelerated in opposite senses of the same direction parallel to X_o.

b) Once reached the speed u, A and B end their corresponding acceleration, and each of them moves for one hour according to their respective identical clocks, at the same uniform speed u in the same direction but in opposite senses with respect to the robot C.

c) Therefore, for one hour each robot remains in an inertial reference frame from whose perspective the other robot moves at a uniform velocity v.

d) During that hour, and only during that hour, A and B register in an appropriate physical support (the same in both cases) the successive ticks of their corresponding clocks, one tick per second, each tick recorded as a short beep.

e) After the programmed hour, A and B accelerate and then de-celerated in such a way that at the instant t_{o1} of RF_o they recover their initial condition of being at rest in RF_o together with the robot C.

Once at rest in RF_o, the recordings of A and B are compared (Figure 13.1). And when comparing the recordings of the beeps produced by A's clock and B's clock, one of the following alternatives will occur:

(a) The recordings do not match: in one of them the time inter-val between any two successive beeps is greater than in the corresponding beeps of the other.

(b) Both recordings match: the beeps are separated by the same time intervals between any two successive beeps.

What cannot happen is that the beeps recorded by A are at the same time less temporally spaced and more temporally spaced than those recorded by B simply because this possibility goes against the Second Law of Logic. By the way, this possibility is at the core of the problem posed by H. Dingle cited above.

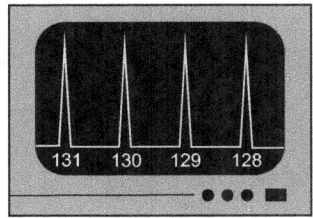 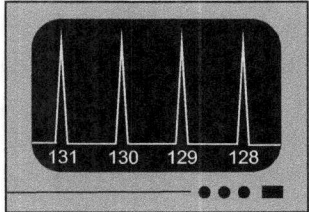

Figure 13.1 – Recordings of the twin robots: one beep per second during one hour of their respective proper time.

We must, therefore, consider only the above two alternatives. But before, recall that according to SR, for the robot A, the clock of B runs slow than its proper clock, and vice versa. Or in other words: robot A's clock is and is not further ahead than robot B's clock; and robot B's clock is and is not further ahead than robot A's clock. And recall that both clocks have been registering their respective beeps while they were moving at a uniform velocity v relative to each other.

In the case of the alternative (a), the symmetry of SR would not hold, and not all inertial reference frame would be equivalent, which goes against the Principle of Relativity. In the case of the alternative (b) the dilation of time observed from different inertial reference frames would only be apparent, as apparent as the refractive deformation of a rod partially submerged in water.

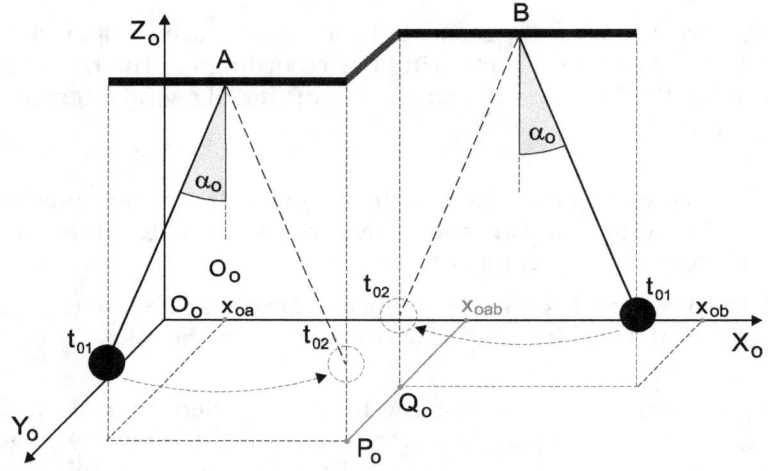

Figure 13.2 – Two identical pendulums A and B in their proper reference frame RF_o just at the instant t_{o1} when they begin to swing.

13.6 Relativistic local simultaneity

The scenario of the argument developed in this section is represented in Figure 13.2. In that scenario, two identical (and ideal) pendulums oscillate in the same conditions in planes parallel to the plane $X_o Z_o$ of their proper reference frame RF_o. And they oscillate as follows:

1. At the instant t_{o1} of RF_o both pendulums begin to oscillate simultaneously and in the same conditions defined by the angle α_o.

2. The ball of the pendulum A begins to oscillate from its initial position whose coordinate on the axis X_o is x_{oa}. This first oscillation of A is to the right (the increasing direction of X_o).

3. The ball of the pendulum B begins to oscillate from its initial

position whose coordinate on the axis X_o is x_{ob}. This first oscillation of B is to the left (the decreasing direction of X_o).

4. At the instant t_{o2} of RF_o both balls reach the end of their first oscillation, which in each case occurs at a different point P_o and Q_o, although both points have the same coordinate x_{oab} on the X_o axis.

5. Both pendulums have the same amplitude of oscillation:

$$x_{oab} - x_{oa} = x_{oab} - x_{ob} \tag{3}$$

6. At the instant t_{o2} both pendulums start their second oscillation: the pendulum A to the left, and the pendulum B to the right, until reaching again their initial positions, and then they begin their third oscillation analogous to the first one.

7. According to the laws of mechanics, swings to the right take the same time for both pendulums as swings to the left.

8. The oscillations of both pendulums are repeated a certain number of times with a certain frequency.

We will now examine the oscillations of the two pendulums A and B from the perspective of the inertial reference frame RF_v which, as always in this book, coincides at a certain instant with the reference frame RF_o (the proper reference frame of both pendulums).

From the perspective of RF_v, the frame RF_o moves with speed $v = kc$, $(0 < k < 1)$, parallel to the X_v axis, in the direction of its increasing values. It is important to note that both RF_o and RF_v are two inertial reference frames to which special relativity can be applied, even though the movements of the pendulums are not uniform. The argument that follows is just a consequence of the Lorentz Transformation, whatever the particular implications of SR and GR may have.

Since in RF_o the two identical pendulums A and B start to oscillate simultaneously at the instant t_{o1} and in the same conditions (angle α_o), in the reference frame RF_v (Figure 13.3), and according to LT, they also oscillate in the same conditions, now defined by the angle α_v, which is related to the proper angle α_o according to:

$$\tan \alpha_v = \gamma^{-1} \tan \alpha_o \tag{4}$$

but they do not begin to swing simultaneously: A starts a time δt_{v1} before B, because when in RF_o both pendulums start to oscillate they are separated in the direction of the relative motion by a proper distance d_o given by (Figure 13.2):

$$d_o = x_{ob} - x_{oa} \tag{5}$$

$$= 2L_o \sin \alpha_o \tag{6}$$

where L_o is the proper length of each of the two pendulums. Consequently, the pendulum A begins a time δt_{v1} before the pendulum B, which is deduced from the Lorentz Transformation and is given by:

$$\delta t_{v1} = \frac{k(x_{ob} - x_{oa})}{c\sqrt{1 - k^2}} \tag{7}$$

Accordingly, the observers of RF_v will have to describe the oscillations of the pendulums A and B as follows:

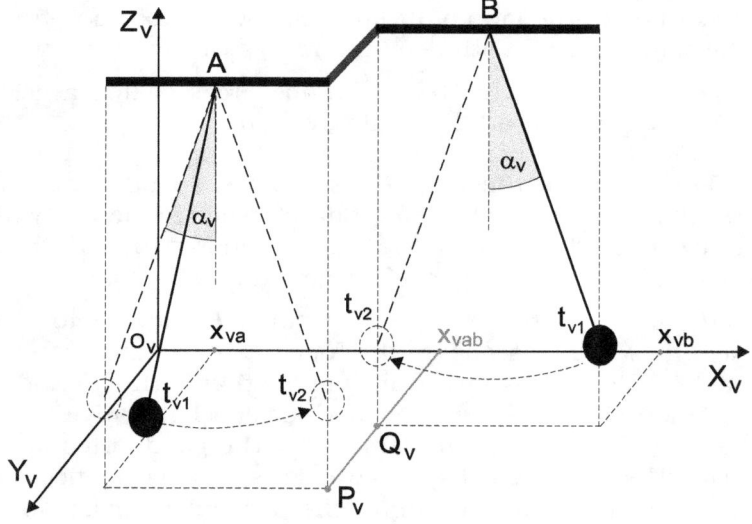

Figure 13.3 – The pendulums A and B from the perspective of the reference frame RF_v.

1. The pendulum A starts its first oscillation a time δt_{v1} before the pendulum B.

2. When the pendulum B starts its first oscillation, the pendulum A has already covered part of its first oscillation.

3. Both pendulums end their first oscillation at the same instant t_{v2} and at two points P_v and Q_v with the same coordinate x_{vab}, because the distance in the direction of the relative motion

between the points P_o and Q_o of RF_o where the pendulums A and B respectively end their first oscillation is zero.

4. Both pendulums have the same amplitude of oscillation:

$$\sqrt{1-k^2}(x_{oab} - x_{oa}) = \sqrt{1-k^2}(x_{oab} - x_{ob}) \tag{8}$$

5. Both pendulums start their second oscillation at the same instant t_{v2}.

6. The pendulum A finishes its second oscillation a time $\delta t_{v2} = \delta t_{v1}$ before the pendulum B.

7. When the pendulum A moves from left to right, it moves slower than the pendulum B.

8. When the pendulum A moves from right to left, it moves faster than the pendulum B.

9. The length of both pendulums first increase and then decrease along each complete oscillation because only in the vertical position $L_v = L_o$, at any other angle $L_v < L_o$ because its component on the X_v axis is not zero and then it is contracted by the factor γ^{-1}.

Thus, according to the observers in RF_v, i.e. according to the Lorentz Transformation:

1. Both pendulums swing slower from left to right than from right to left

This consequence deduced from the Lorentz Transformation is contradictory to all the theoretical and empirical knowledge about the oscillation of pendulums.

13.7 Consequences of the relativistic time deformations

In the case of time dilation, within the framework of H. Ives' theory of absolute space and time, the contradiction that occurs in the case of relativistic time dilation does not occur, when the dilation is considered real instead of apparent. Therefore, in the case of H. Ives it could be a real dilation, or the consequence of a possible anisotropy in the speed of light.

In the case of special relativity, and in accordance with the previous two sections, relativistic time dilation and local relativistic simultaneity imply some consequences incompatible with some of the fundamentals of logic and mechanics:

1. Given any two identical clocks, one of them is and is not further ahead than the other (robots argument).

2. A freely swinging pendulum swings faster in one direction than in the opposite direction without any force other than gravity acts on the pendulum mass (pendulums argument).

The solution of both incompatibilities could be that the relativistic deformations of time are only apparent. Therefore, the special theory of relativity is either inconsistent or describes deformations that are not real but apparent, as is the case with refractive deformations.

13.8 Time in CALMs

Introducing CALMs as a model to explain the physical world does not solve any of the limitations imposed on human knowledge by the (potentially) infinite regress of definitions, arguments and causes. Limitations that are unavoidable for human knowledge, but which are rarely echoed in written or spoken human knowledge; as they do, with some pomp, of other much narrower and more debatable limitations occasioned by the debatable self-reference [194]. In the article 8 of this series, the inevitability of the limitations imposed by the Aristotelian infinite regress of arguments was demonstrated, and in this series of articles it has been extended to definitions and causes:

> Any recursive sequence S of proofs, definitions or causes in which there is a last element to be proved (defined, caused) and each element has an immediate predecessor that proves (defines or causes) it, is uncompletable.

Changing the reference model (the infinite continuum for the finite discreteness) to explain the world does not mean overcoming these limitations. We still need to use primitive concepts. We should not forget this inevitable limitation of human knowledge.

Whether continuous or discrete, the concept of time is surely one of those basic indefinable, primitive concepts. What can be affirmed is that as opposed to the infinitist time continuum modeled by $\mathbb{R}^+$, time in a CALM is always finite and discrete, with minimal indivisible units that here we are calling qutits. The qutits also present adjacency (contiguity, immediate successiveness): each qutit is immediately followed by another qutit without any other qutit existing between them, in the same way that every natural number n is followed by another natural number $n + 1$, without any natural number existing between n and $n + 1$.

Recall that a CALM could exist in a sequence of alternating states: the perceptible *permanence mode*, in which the state of its qusits (discrete space units) does not change; and the imperceptible *interacting mode*, which is simultaneous with the permanence mode, although it remains in an imperceptible background (such as computer applications running in the background) while changing the state of each qusit for the next qutit, according to the CALM's rules.

Unlike instants (which have no duration), qutits do have a duration greater than zero, all of them the same. Thus, in a CALM time is not a densely ordered sequence of instants without duration. In a CALM time is a strictly ordered (with immediate successiveness) and finite sequence of immediately successive qutits. And a time interval will always be defined by a finite number of successive and adjacent qutits. In a CALM time is therefore absolute. And since space is also absolute (article 12 of this series), only absolute motion is possible in CALMs. Although in the physical world (whether or not modeled by a CALM) preinertia makes it impossible to detect and measure absolute motion. Only relative motion resulting from different absolute motions of physical objects through the real physical space can be observed and measured. Except, perhaps, in the conditions analyzed in the Santiago del Collado experiment.

Paper 14. Relativity and discreteness

Abstract. Leaving aside the fact that special relativity is a theory built on the infinitist spacetime continuum, and that this space-time continuum is inconsistent (see paper 3 of this series), here it will be proved that special relativity is not compatible with discrete space and time. Therefore, the relativistic inertial length contractions, inertial time dilations and local simultaneities could be only apparent, as is apparent the continuous motion in a film, or the refractive deformations; or even continuous motion in a discrete universe. It also happens that the relativistic Lorentz factor, which is crucial in the Lorentz Transformation, coincides with the conversion factor between the continuous and the discrete versions of Pythagoras Theorem.

Keywords: special relativity, Lorentz factor, discrete space and time, Discrete Pythagoras Theorem.

14.1 Introduction

Although the main purpose of this article is to formally demonstrate the incompatibility of the special theory of relativity (SR) with a discrete space and time scenario, we will make a couple of preliminary considerations about SR. The first has to do with the excessive weight that, in my opinion, SR has in contemporary physics: SR seems to have become a new basic law of logic: if something does not agree with SR, that something, whatever it is, cannot be true. The other consideration has to do with the experimental support of the theory. Indeed, it is often forgotten that:

1. Experimental confirmations of SR (almost always non-inertial (!) time dilations) can also be confirming other theories. For example, of the theory of absolute space and time of H. E. Ives

185

([163, 157, 156, 158].

2. Some of the observed relativistic spacetime deformations may be only apparent, as are all refractive deformations.

3. Or they could be consequences of considering as continuous a spacetime that in fact is discontinuous, discrete. The last section of this paper proves the factor that converts between the continuous and the discrete version of Pythagoras Theorem is just the relativistic Lorentz factor.

4. The confirmations have to be simultaneously symmetric. Indeed, let A and B be two inertial reference frames in relative motion:

   - In the direction of relative motion, the proper objects of the reference frame A are seen shorter from the frame B than from the frame A. And, at the same time, the proper objects of the reference frame B are seen shorter from the frame A than from the frame B.

   - The proper clocks of the reference frame A are observed slower from the frame B than from the frame A. And, at the same time, the proper clocks of the reference frame B are observed slower from the frame A than from frame B.

   - In the direction of relative motion, all events that are simultaneous in the reference frame A are not simultaneous in the reference frame B. And all events that are simultaneous in the reference frame B are not simultaneous in the reference frame A.

   - In all cases, SR requires both confirmations, but that never happens.

It is also surprising that the relativistic contractions are the same for all objects, regardless of their chemical composition and crystallographic structure. And that inertial time dilation is independent of the periodic mechanisms by which the different clocks measure time. Moreover, there is no physical cause that produces these deformations, only the fact that they are observed with different uniform relative velocities.

14.2 Special relativity is not compatible with discreteness

According to the Principle of Relativity, the laws of physics are the same in all inertial reference frame. And according to the Theorem of the Consistent Universe 5.1 the universe evolves under the control of a unique set of invariant and formally consistent laws. This includes all the universal constants involved in such laws. So, being Planck length l_p, and Planck time t_p two universal constants, they should be universal constants in all inertial reference frames, which poses the problem of their respective relativistic contraction and dilation. Or in other words, the problem of the relativity of the intervals of space and time below which physical laws do not apply, will now depend on the relative velocity at which the corresponding events and magnitudes are observed and measured. This problem has already been dealt with by some authors, although they have not had a great impact [11, 164, 5, 4, 150].

It could be argued that the Lorentz Transformation does not hold for lengths and times respectively less than l_p and t_p. So, let L_o be the length of a macroscopic rule in its proper inertial reference frame RF_o. In a discrete space and time, and according to the Theorem 5.6, we will have:

$$L_o = n_o l_p; \ n_o \in \mathbb{N} \tag{1}$$

Let RF_v be another inertial reference frame that coincides with RF_o at a certain instant and from whose perspective RF_o moves with a uniform velocity $v = kc$, $0 < k < 1$, in the direction of the increasing X_v. In accordance with the Lorentz Transformation, the rule, that moves parallel to X_v, will be observed with a length L_v such that:

$$L_v = \gamma^{-1} L_o \tag{2}$$

where $\gamma = 1/(\sqrt{1 - k^2}$ is Lorentz factor. If l_p is also a universal constant in RF_v, we will have:

$$L_v = n_v l_p; \ n_v \in \mathbb{N} \tag{3}$$

In consequence, it must hold:

$$n_v l_p = \gamma^{-1} L_o = \gamma^{-1} n_o l_p \tag{4}$$

$$n_v l_p = \gamma^{-1} n_o l_p \tag{5}$$

$$n_v = \sqrt{1 - k^2} \, n_o \tag{6}$$

which is impossible because $\sqrt{1 - k^2}$ is not a natural number. The same argument applied to any proper interval of time $t_o > t_p$ leads to:

$$t_v = t_o/\sqrt{1-k^2} \qquad (7)$$

$$n_v t_p = n_o t_p/\sqrt{1-k^2} \qquad (8)$$

$$n_v = n_o/\sqrt{1-k^2} \qquad (9)$$

which for the same above reason is also impossible. If l_p and t_p represent respectively the minimum indivisible unit of length and time, we have to conclude that the theory of special relativity is not compatible with a discrete space and a discrete time, i.e. it is not compatible with the Theorem of the Discrete Threshold 5.8. Therefore, the special theory of relativity requires that one of the following two alternatives be satisfied:

1. The laws of physics hold for any space (time) interval arbitrarily less than the discrete space (time) unit.

2. The speed of light is undefined for any time interval and any length interval respectively different of nt_p, and nl_p, for any natural number n.

14.3 Pythagoras theorem and Lorentz factor

(The text of this section is taken from [197, pp. 438-440])

If, according to Theorem 3.8 of the Inconsistent Continuum, the continuum is inconsistent, then the only alternative to the spacetime continuum would be a discontinuous, i.e. discrete, space and time made of indivisible units (atoms of space and time in L. Smolin words [304]) we are calling here respectively qusits and qutits. The interest in discrete space and time began in the first half of the twentieth century [31, 61], although only in a minority of authors. W. Heisenberg, for instance, considered the idea of space as a sort of crystal lattice composed on tiny cells of the size of an elementary particle, although this idea was not finally developed [99]. Things have begin to change, especially in the last two decades. [108, 225, 176, 112, 22, 279, 23, 17, 274, etc.].

An increasing number of physicists suspect now that, in fact, Planck length and Planck time define a sort of space and time 'granularity' that could be an efficient alternative to the infinitist spacetime continuum, an alternative that could be tested in experimental terms [236, 63, 109, 200, 54]. The discrete nature of space and time has been proposed in different areas of physics

[165, 124, 325, 100, 303, 304, 312, 20, 203, 211, 9, etc.], albeit the proposed models continue to be developed within the framework of infinitist mathematics.

Although discrete geometries already exist, they exist for particular purposes, for example the combinatorial analysis of the relationships between geometric elements [29], or the development of computational algorithms for the representation of geometric objects [70, 53]. There are even general discrete geometries, whether or not applied to quantum gravity, but not independent of infinitist mathematics. The discrete geometry suggested here would be a geometry with extended and indivisible minimum units of space instead of the non-extended points of current geometries. A discrete geometry that could only be developed on a discrete and finitist basis. For this geometry, everything remains to be done, starting with the establishment of its foundational base (axioms and definitions). Even so, some non-detailed arguments, as the next one, can be made.

Though the fine structure of (a possible) discrete space is unknown, let us consider the right angled triangles depicted in Figure 14.1.

It can easily be tested that the number of qusits (in reality bidimensional qusits) of their corresponding hypotenuses is in each case equal to the number of qusits of the greater of their corresponding legs. This empirical version of discrete Pythagoras theorem could surely be formally proved once the foundation of discrete geometry had been formally established. If that were the case, the factor for converting between continuous and discontinuous hypotenuses would have the algebraic form of the relativistic factor γ.

Indeed, let h, x and y be the respective number of qusits of the hypotenuse and legs of a right triangle in a discrete space time, and let l_p be the length of one qusit in both the discrete and the continuous geometry. Assume $x < y$. In a discrete geometry we will have: $h = y$. In the continuous geometry the length of the hypotenuse would no longer be hl_p but $h'l_p$, being $h' > h$, because it is greater than the length hl_p, which is also the length yl_p of the greatest leg (note that while h, x and y are natural numbers, l_p and h' are real numbers). According to classical Pythagoras theorem, it can be written:

$$\text{Hypotenuse: } h'l_p = \sqrt{(yl_p)^2 + (xl_p)^2} \qquad (10)$$

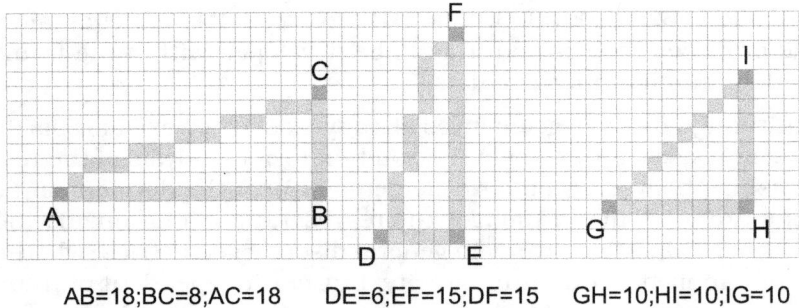

AB=18;BC=8;AC=18   DE=6;EF=15;DF=15   GH=10;HI=10;IG=10

Figure 14.1 – A crude representation of three right angled triangles to test Pythagoras discrete theorem. The size of bi-dimensional space absolute quanta (qusits) has been exaggerated to makes it possible their visual counting.

$$\text{leg: } yl_p = \sqrt{(h'l_p)^2 - (xl_p)^2} \tag{11}$$

$$y = \sqrt{h'^2 - x^2} \tag{12}$$

The ratio between the continuous and the discrete hypotenuse is given by:

$$\frac{h'l_p}{hl_p} = \frac{h'}{h} \tag{13}$$

$$= \frac{h'}{y} \tag{14}$$

$$= \frac{h'}{\sqrt{h'^2 - x^2}} \tag{15}$$

$$= \frac{1}{\sqrt{1 - (x/h')^2}} \tag{16}$$

where the last term on the right side of (16) as the algebraic form of the relativistic Lorentz factor γ. It can ve rewritten as:

$$\frac{h'l_p}{hl_p} = \frac{1}{\sqrt{1 - (xl_p/h'l_p)^2}} \tag{17}$$

Let a^* be a photon that moves through a vertical distance yl_p in the rest frame RF_o of its source. Assume a^* moves the same vertical distance yl_p from the perspective of another inertial frame RF_v while RF_o moves with respect to RF_v the horizontal distance xl_p

at a uniform velocity v parallel to X_v for a time t_v. So, a^* moves with respect to RF_v along the hypotenuse of a right triangle whose legs are yl_p and $xl_p = vt_v$; i.e. a^* moves along the hypotenuse $h'l_p$ (10). And it will hold $h'l_p = ct_v$. Therefore, (17) can be rewritten:

$$\frac{h'l_p}{hl_p} = \frac{1}{\sqrt{1 - (vt_v/ct_v)^2}} \tag{18}$$

$$= \frac{1}{\sqrt{1 - (v/c)^2}} \tag{19}$$

$$= \gamma \tag{20}$$

which proves the ratio between the continuous hypotenuse and its corresponding discrete alternative is the relativistic Lorentz factor γ. This result suggests that a discrete interpretation of special relativity could be possible. Special relativity could be, in fact, the consequence of explaining a discrete, discontinuous, world in terms of the continuous mathematics of the spacetime continuum. Or in other more expeditious words: the result of explaining a consistent discontinuous reality in terms of an inconsistent continuous reality.

Paper 15. The origin of the universe

Abstract.-This article discusses one of the most fundamental problems in cosmology, and therefore of this series of articles: the origin of the universe. The Theorem 5.8 of the Discrete Threshold makes the laws of physics inoperable in the first discrete interval of time of the history of the universe. Although for other reasons, the same conclusion holds for the first discrete (quantum) time interval $(0, t_p)$ defined by the Planck time t_p. So, contemporary physics can only describe the history of the universe after that first lapse of time. Hence, the problem posed by the origin of the universe cannot be solved with the sole help of the physical laws. And, in fact, all the solutions proposed in the framework of contemporary physics are pseudo-solutions or evasions of the problem. This article demonstrates that this is the case. In fact, the only consistent way to approach the problem of the origin of the universe is to deal with the infinite regress of causes and its formal consequences, demonstrated in article 8 of this series of articles. This confrontation leads to a unique solution that will have to be approached without religious or anti-religious prejudices.

Keywords: Big Bang theory, Theorem of the Consistent Universe, Theorem of Formal Dependence, Theorem of Discrete Threshold, flat universe, finite universe, irreversible time, Augustine of Hippo, Kant antinomy, infinite regress of causes, first cause.

15.1 Introduction

After a brief introduction to the Big Bang theory, this article recalls that the universe is consistent and time is irreversible (a result that was demonstrated in paper 10 of the series). The The-

193

orem 5.6 of Formal Dependence and the Theorem 5.8 of Discrete Threshold are then recalled, and they will be of paramount importance in the subsequent discussions. Next, a fact that seems convenient to take into account when reflecting on the origin of the universe is examined: the critical energy density of the universe, which makes it a flat universe capable of evolving in the same direction for billions of years.

Afterwards, the reflections of Augustine of Hippo on the origin of time are recalled, which serves as an introduction to the analysis of the pseudo-solutions proposed by contemporary physics to the problem of the origin of the universe. It is then demonstrated that, in effect, they are not true solutions. For contemporary physics, the problem of the origin of the universe remains a problem. And it will continue to be so as long as we try to solve it with the sole help of the physical laws, simply because the existence of the physical laws cannot be confirmed during the first time interval $[0, t_p)$, however short the Planck time $t_p = 5.39124x10^{-44}$ seconds (or any other first discrete time interval), an inevitable consequence of the Theorem 5.8 of Discrete Threshold.

At this point, it only remains to recall the formal results obtained in paper 8 on the infinite regress of demonstrations, definitions and causes. And to deal with the infinite regress of causes, which might be the only way to deal with the problem of the origin of the universe. Theorem 8.2 of the First Element tells us that also for the universe there should exist a first cause external to the universe itself. This implies that this first cause is inexplicable in terms of other causes, because if it were explicable in terms of other causes it would cease to be the first cause, in which case we would either enter into a potentially infinite regress of successive causes, or into a circular explanation of causes; both conclusions inadmissible from the formal point of view.

15.2 The Big-Bang theory

Most cultures have had their own theory about the origin of the universe, almost always related to entanglements or whims of gods, even to some divine vomit [135, p. 141]. The most rational theories come, like almost all theories of this type, from the Greek world. The simplest and least compromising way, physically and metaphysically, was to assume, as Aristotle did, that the universe was eternal and unchanging, that it always existed as we now see it. In this way it was not necessary to give more explanations neither about its origin, nor about its evolution.

The essence of the Aristotelian cosmogony (of an eternal and stationary universe) was maintained until well into the 20th century, when certain key discoveries such as the expansion of galaxies, fossil radiation and the abundance of chemical elements, led scientists to elaborate a cosmogony radically different from the Aristotelian theory: the Big-Bang theory, whose name (jokingly chosen) is due to one of its opponents: Fred Hoyle (defender of the stationary universe [151]).

The Big-Bang theory [113, 114, 183, 184, 185, 186, 187] explains the evolution of the universe reasonably well, except for the formation of the universe itself, i.e. except its origin. It is a theory with a wide and diverse experimental support. It is also compatible with the Principle of Directional Evolution, the only inductive principle assumed in this series of articles, and therefore with all the theorems deduced from them. Particularly with the Theorem 5.8 of the Discrete Threshold, according to which the physical laws do not apply in distances less than the discrete (quantum) unit of length and time, (surely defined by Planck length and Planck time).

The Big-Bang theory is (almost) unanimously accepted in our days and it is widely explained in many books on different aspect of physical reality ([311, 179, 314, 268, 339, 178, 268, 282, 101, 137, 135, 136, 145] etc.), notwithstanding some aspects of the theory are still under discussion. One of the strangest stages of the early evolution of the universe is the so-called inflation stage, in which space (real or unreal?) would have expanded in a super-exponential way. But if space is not real, what is it that expanded in such an enormous way? From the point of view of CALMs, things would have been simpler, without the need for any spatial inflation, as will be seen below.

The consistency of the universe, on the other hand, implies that all the consequences derived from the laws of logic must be assumed, including its modes of inference and the infinite regress of demonstrations, definitions and causes. And it is exactly what happens in the study of the components, properties and evolution of the universe. It only remains to apply them also to the problem of its origin. Because the logical analysis of a consistent universe also includes the logical analysis of its origin, which implies facing the problem of the infinite regress of causes, which, as will be seen, has a single solution (already announced in Theorems 8.2 and 8.1), although this is a very difficult solution to deal with due to religious and anti-religious prejudices.

15.3 The universe is consistent and time is irreversible

All the experimental scientific knowledge accumulated over the last 300 years points to the same inductive conclusion: the universe has been evolving for its entire history (over 13.8 billion years) under the control of an unchanging set of formally consistent laws. Where formally consistent means that each of these laws, under the same conditions, always leads to the same type of results. This is also what the Theorem 5.3 of the Consistent Universe Theorem states. No exception to the consistent behavior of the Universe is known: things and processes are what they are rationally expected to be, although in some things and procedures what is rationally expected seems stranger than what is intuitively expected. Its universal consistency is the reason why the universe allows itself to be described with mathematical languages and also with computational languages (perhaps more appropriate than mathematical languages). And the reason for which it is absolutely reasonable to assume the Principle of Directional Evolution introduced in paper 5 of this series, which includes the independence from rational observers because for most of its history there were no rational observers in the universe:

> **Principle 5.1 of Directional Evolution**: *The universe always evolves independently of its rational observers and in the same direction of increasing its global entropy.*

from which it immediately follows the Theorem 5.3 of the Consistent Universe proved in the same paper 5:

> **Theorem 5.3 of the Consistent Universe**: *The universe evolves under the control of a unique set of invariant and formally consistent laws.*

Although it is a formal theorem, its empirical evidence is possibly the greatest of all the inductive evidences established in science:

> No inconsistent natural fact has ever been observed in the physical world.

The Theorem 5.5 of Identicality also proved in article 5 is also of enormous empirical evidence:

Theorem 5.5 of Identicality: *All particles of the same type have the same properties at rest.*

From this theorem two corollaries follow immediately, and these corollaries allowed to prove in paper 10 the Theorem 10.2 that states te existence of an arrow of time, i.e. the irreversible nature of time:

Theorem 10.2 of the Arrow of Time: *In a consistent universe the joint evolution of any system and its environment is always in the same direction of increasing its entropy.*

Therefore, our universe must have begun in a state of maximum anisotropy, which is compatible with the Big-Bang theory that describes the evolution of the universe once it originated, although not the origin itself. Furthermore, the laws that explain the evolution of the universe cannot explain the very origin of the universe, if, as it seems, the Theorem 5.8 of the Discrete Threshold is also verified:

Theorem 5.8 of the Discrete Threshold: *The laws of physics do not apply in spaces smaller than the indivisible unit of space nor in times smaller than the indivisible unit of time, both being of non-zero extension.*

As we will se in this paper, the only way of facing the problem of the origin of the universe is to make use of the Theorem 5.6 of the Formal Dependence:

Theorem 5.6 of the Formal Dependence: *No concept defines itself; no statement proves itself; no physical object is the cause of itself; and no cause is the cause of itself.*

and theorems deduced from it regarding the infinite regress of causes.

15.4 The universe is flat

As is well known, the general theory of relativity states that, depending on its energy density at the time of its formation, the universe could be closed (elliptical geometry with curvature greater than zero), open (hyperbolic geometry of curvature less than zero) and flat (Euclidean geometry of zero curvature). In the first two cases, the al-Tutsi-Legendre version of Euclid's 5th Postulate [193] is not verified (in the elliptic case the internal angles of a triangle add up to more than 180° and in the hyperbolic case less than 180°). Also in the case of the closed universe, the universe would collapse gravitationally, while in the case of the open universe a non-decreasing expansion would continue forever [313, 101, 216, 269, 37, 84, 81, 82, 78, 80, 335, 206, 137, 121, 268, 311, 314, 179, 36].

There is a single value for the initial energy density of the universe that separates closed universes from open universes, the value that corresponds to the flat universe. A universe that, after an accelerated expansion, will continue to expand indefinitely but asymptotically approaching zero. This unique density is called the critical density ρ_{crit}. The energy density of the present universe has been calculated from astronomical observations in three independent calculations: the energy density due to ordinary matter ($\rho_{om} = 0.049\rho_{crit}$), the energy density due to dark matter ($\rho_{dm} = 0.268\rho_{crit}$) and the energy density of dark energy ($\rho_{de} = 0.683\rho_{crit}$). As can be seen, the sum of the three independent measurements is precisely the critical density ρ_{crit}, which corresponds to that of the flat universe:

$$\rho_{om} + \rho_{dm} + \rho_{de} = \rho_{crit} \tag{1}$$

It should be noted that in the case of a closed universe and in the case of an open universe, the formation of structures such as galaxies would be compromised, at least in the long and very long term. And it should also be noted that the initial energy of the universe had to be such that its initial energy density could not differ from the critical density ρ_{crit} by a factor greater than 10^{-62}, a extremely small factor. Explaining this coincidence is one of the biggest problems facing cosmology today. And as will be seen in the next section, it also has a great interest in the discussion about the origin of the universe. Obviously, equation (1) can be considered, or not, as a mere random coincidence.

15.5 Going back to Augustine of Hippo

Before starting our formal analysis of the origin of the universe, we will recall a similar (and well-known) problem that Augustine of Hippo (Saint Augustine) faced as early as the 4th century AC, and which is reflected in the following quote (often manipulated and misrepresented) [67, Book XI, p.581]:

> And I will stop, and I will solidify in you, in my mold, your Truth, and I will not bear the questions of people who, sick of their own punishment, are thirstier than they can and say: What was God doing before He made heaven and earth? or What came to his mind to do something, when he had never done anything before? Give them, Lord, the ability to think carefully about what they say and discover that "never" cannot be said where there is no time. Therefore, of whom it is said that he "had never done," what else is said but that he did it out of time? See, therefore, that it is impossible for there to be any time without creation and stop saying such vanity. May they also project themselves towards all that is before and conceive of you as the eternal creator of all times before all times, and that no time is alien to you, nor any creation, no matter how far beyond time it may be.

Note that, as is frequently said [339, p. 162 Kindle]:

> The question: What happened before God created the universe? is meaningless.

is not what St. Augustine said:

> May they also project themselves towards all that is before and conceive of you as the eternal creator of all times before all times, and that no time is alien to you, nor any creation, no matter how far beyond time it may be.

And returning to the modern literature on the matter [339, p. 162 Kindle Edition]

> Essentially, Augustine's answer survives its translation into the language of modern physical cosmology. Nothing precedes the origin of the universe because, in that context, time -that which clocks measure- has no meaning.

As will be seen in the next section, the above words reflect a widespread attitude among contemporary physicists. Actually, before the origin of the observable universe, the observable universe did not exist. And something must have happened for it to come into existence. Facing this problem requires stripping oneself of all kinds of prejudices, religious and anti-religious.

15.6 Was the universe created?

As noted above, the Big-Bang theory is a theory about the evolution of the universe, not about its origin. It is a theory that explains reasonably well the evolution of the universe from the instant $t_p = 5.39124x10^{-44}$ seconds until our days, but it does not explain what could have happened before. And the reason is quite simple: according to the Theorem 5.8 of Discrete Threshold the laws of physics cannot be applied in intervals of time less than the discrete time interval $[0, t_d)$ (maybe Planck time $t_d = t_p$), which includes the instant $t = 0$.

Let us begin by posing the problem of the origin of the universe in terms of the Kantian antinomy:

A. The world had a beginning in space and is limited in time.

B. The world is infinite in both space and time.

Therefore, and taking into account Corollary 3.3 (that establishes the inconsistency of the actual infinity) if the world is consistent then it had a beginning and is limited in time. Until now, of course, physics has not taken Corollary 3.3 into account. And when it has tried to solve the problem of the origin of the universe, what he has been really doing is postponing it, camouflaging it, hiding it, diluting it. Among the most common of these false solutions are the following:

1. The Universe does not have an origin because it is eternal.

   Objection: if it is eternal, it is infinite and contains infinite ω-ordered sequences of instants. For example, from this very moment we can count the seconds passed backwards in time: 1s ago, 2s ago, 3s ago,... and we would have an ω-ordered sequence of seconds already passed. And according to Corollary 3.2, it would be an inconsistent sequence. Therefore an eternal universe would be inconsistent.

2. The universe is created and annihilated successively, in an infinite number of cycles in which each cycle begins with a Big-Bang, which is followed by an expansive stage, then a contractive stage and a new Big-Bang (Big-Crunch).

Objection: the same as in the previous case, now applied to the infinite number of cycles. We would have another case of inconsistent universe.

3. It makes no sense to ask what was there before t = 0, just as it makes no sense to ask what is further north of the earth's north pole on any terrestrial path leading to the earth's north pole.

   Objection: Actually, it does make sense: further north of the Earth's north pole begins a terrestrial path towards the Earth's south pole. But here we are not interested in what happened *before* $t = 0$, but in what happened in the interval $[0, t_p)$, which includes what happened at the precise instant $t = 0$. *We are interested here in what happened just at the instant in which the 'north pole $t = 0$' is reached*, and in the interval of time that immediately succeeded it whose duration is the indivisible unit of discrete time t_d (Theorems 5.10 and 5.11)

4. The most widespread pseudo-answer (by far) is the claim that the universe originated from a fluctuation of nothingness. The entire universe with its endowment of formally consistent laws (Theorem of the Consistent Universe) and with just the right energy density to ensure its directional evolution for billions of years (flat universe) arose from a fluctuation of nothingness. What a fluctuation!

   Objection: If the universe originated from a fluctuation of nothingness, then nothingness would have physical capabilities, at least the capability to fluctuate universes. But something that has physical capabilities is not nothingness, it is something that has physical capabilities and can be defined in terms of those physical capabilities. Nothingness cannot have capabilities. Otherwise, nothingness could be something. So, the problem of the origin of the universe now becomes the origin of that something capable of originate universes. The same problem with different objects. Indeed, if the formal rigor of language is eliminated, we will be able to demonstrate whatever we want.

Thus, no matter how many fluctuations are prescribed, the problem of the origin of the universe remains unsolved. This should not be a surprise if we take into account the formal elements involved, which are worth remembering at this point:

1. **Principle 5.1 of Directional Evolution**: *The universe always evolves independently of its rational observers and in the same*

direction of increasing its global entropy.

2. **Theorem 5.3 of the Consistent Universe**: *The universe evolves under the control of a unique set of invariant and formally consistent laws.*

3. **Theorem 5.6 of Formal Dependence**: *No concept defines itself; no statement proves itself; no physical object is the cause of itself; and no cause is the cause of itself.*

4. **Theorem 8.1 of the Uncompletable Regress**: *Any recursive sequence S of proofs, definitions or causes in which there is a last element to be proved (defined, caused) and each element has an immediate predecessor that proves (defines or causes) it, is uncompletable.*

5. **Theorem 8.2 of the First Element**: *A sequence in which there is a last element and each element has an immediate predecessor is a complete totality only if it has a first element.*

Infinite regress applies to demonstrations, definitions and causes. In the case of demonstrations we have to make use of axioms and basic principles. In the case of definitions we have to admit the use of primitive, undefined concepts. But what to do in the case of causes? It seems inevitable to think of a first external cause as the origin of the universe. As noted above:

> This implies that this first cause is inexplicable in terms of other causes, because if it were explicable in terms of other causes it would cease to be the first cause, in which case we would either enter into a potentially infinite regress of successive causes, or into a circular explanation of causes; both conclusions inadmissible from the formal point of view.

We shall finally have to admit a first cause external to the universe as the origin of the universe, the origin of all that has come after. That is to say:

> Taking into account the formal consistency of the universe and the exquisite adjustment of its initial energy density, which ensure its consistent evolution in the same direction over billions of years, it is not unreasonable to think that the universe was intentionally created by a system outside the universe itself. And since all our logic and all our knowledge can only be

applied to the objects and processes of our universe, we can know nothing about that system precisely because it is alien to the very universe we inhabit, and in which we have developed all our cognitive faculties and all our inductive knowledge. It would not be formally correct to project our knowledge of this physical world to realities outside this physical world.

In any case, it would be the creation of an extraordinary object that has evolved from a cloud of quarks and gluons to Cervantes' Quixote or Mahler's Symphony Nº 2. And, *above all*, to reasoning, respect and compassion. Although also towards dogmatism, intolerance and violence. And towards the freedom to choose our own personal and social destiny.

Paper 16. A discrete solution to Kant's antinomies

Abstract.-This penultimate article of the series is a kind of parenthesis in which, taking advantage of the results demonstrated in the previous articles, Kant's four antinomies are analyzed. It is shown that in reality they are authentic contradictions, opposite statements formally deduced from a hypothesis and a physical concept that were ignored by Kant and his contemporaries, and that are made explicit here: the Actual Infinite Hypothesis (first, second and fourth antinomies) and the physical information introduced into the universe by living beings (third antinomy). Consequently, and considering the discreteness that follows from the inconsistency of the Axiom of Infinity and of information, Kant's antinomies would be indicating the need for a finitist and discrete model to explain the physical world.

Keywords: antinomy, contradiction, Kant's antinomies, actual infinity, physical information, discrete model.

16.1 Introduction

An antinomy is a pair of apparently incompatible statements such that each of them admits a proof of truth. They would therefore be related to paradoxes and contradictions. Although unlike contradictions, in the case of antinomies there would be no foundational assumption from which both antinomical statements could be formally deduced. If this were the case, it would be a simple contradiction that is demonstrating the inconsistency of the corresponding foundational assumption.

At least in the case of Kant's famous four antinomies, it will be shown here that they are found on implicit assumptions. They are, in fact, contradictions. Three of them would be demonstrating the inconsistency of the Actual Infinity Hypothesis. The fourth would be demonstrating the inconsistency of physical determinism as the only cause of the evolution of the universe. Because, in effect, the universe has managed to produce informed systems that create and execute physical information, understood as the ability to produce arbitrary changes, like having red feathers or yellow feathers, that cannot be deduced from physical laws. Those systems are living beings.

After recalling Kant's four antinomies, I will present the formal instruments for demonstrating their contradictory nature and pointing out the cause of their inconsistency. All these formal instruments have been demonstrated in the previous articles of this series of articles dedicated to propose the necessity of founding a discrete and finitist cosmology. Finally, the contradictory nature of each of Kant's four antinomies will be demonstrated, pointing also to their respective inconsistent implicit assumptions.

It would then be legitimate to consider Kant's antinomies as proof of the inconsistency of the actual infinity, as well as of the strict physical determinism as the only cause of the evolution of the universe. Kant's antinomies would have been announcing for almost three centuries the formal need to consider finitist and discrete cosmologies instead of the classical infinitist models. They would have been advertising the need for this series of articles!

16.2 Kant's four antinomies

Kant defines the concept of antinomy in his Critique of Pure Reason [168, p. 162]. A text of great metaphysical hardness, unkind to the absent-minded reader. Although on the other hand, this serious metaphysical density is as unserious, as apparent, as any other: it is built on primitive concepts (and that is inevitable: not everything can be defined). But, and here is the slip, *it is built on primitive concepts without recognizing that it built on primitive concepts.* For example, in the 196 pages of his Critique of Pure Reason, Kant uses the word "infinite" 56 times without giving a prior definition of that word, "as beings well known to all" as would Newton say [239, p. 77]. And that changes everything.

It would not make sense to reproduce here the Kantian definition of antinomy; to understand it, one would also have to reproduce the text of the 162 pages that precede it. The definition given above, which is also the one that emerges from the statements of the antinomies themselves, is enough for us. Thus, Kant's four antinomies can be stated as follows:

- ANTINOMY 1

  Thesis: The world has a beginning in time and with respect to space it is also enclosed in limits.

  Antithesis: The world has no beginning and no limits in space, it is infinite, both in time and space.

- ANTINOMY 2

  Thesis: Every compound substance in the world is composed of simple parts; and there is nothing but the simple or the compound of the simple.

  Antithesis: No compound thing in the world is made up of simple parts; and there is nothing simple in the world.

- ANTINOMY 3

  Thesis: Causality according to the laws of nature is not the only one from which the phenomena of the world can all be deduced. It is also necessary to admit, in order to explain them, a causality by freedom.

  Antithesis: There is no freedom whatsoever, but everything in the world happens only according to the laws of nature.

- ANTINOMY 4

  Thesis: To the world belongs something which, as its part or as its cause, is an absolutely necessary being.

  Antithesis: There is nowhere an absolutely necessary being, either in the world or outside the world, as its cause.

Note the actual infinity is directly compromised in Antinomy 1; the Axiom of Infinity division is compromised in Antinomy 2; the infinite regress of causes is compromised in Antinomy 4; The non-existence of informed system with arbitrary properties is compromised in Antinomy 3.

16.3 Formal tools to solve Kant's four antinomies

The infinity, the infinite division, the infinite regress and an attribute of a living being, liberty, play a capital role in Kant's antinomies. Obviously, in 1787 (year of the publication of the Critique of Pure Reason) there were neither modern infinitist mathematics nor modern atomic theory, nor quantum mechanics, all of which would be born a little over a century later. But as can be deduced from Kant's own text, the infinity seems to be the actual infinity:

> ... Therefore, in order to think of the world that fills all the spaces as a whole, the successive synthesis of the parts of an infinite world would have to be considered as complete, that is, it would have to be considered as an infinite time elapsed in the enumeration of all existing things. [168, p. 181].

> ... Infinite is a magnitude over which no larger magnitude is possible [168, p. 182]..

In any case, and as it cannot be otherwise, Kant's antinomies will be examined here in accordance with contemporary formal and experimental sciences. In this analysis, the following results will be used, which were already justified, introduced and demonstrated in the previous articles of this series of articles:

> **Principle 5.1 of Directional Evolution**: *The universe always evolves independently of its rational observers and in the same direction of increasing its global entropy.*

> **Theorem 5.3 of the Consistent Universe**: *The universe evolves under the control of a unique set of invariant and formally consistent laws.*

> **Theorem 5.6 of Formal Dependence**: *No concept defines itself; no statement proves itself; no physical object is the cause of itself; and no cause is the cause of itself.*

> **Corollary 3.3 of the Inconsistent Infinity**: *The actual infinity is inconsistent.*

> **Theorem 3.8 of the Inconsistent Continuum**: *The spacetime continuum is inconsistent.*

> **Theorem 3.12 of the Finite Universe**: *A consistent universe cannot contains an actual infinite number of physical objects.*

> **Theorem 5.1 of the Finite Lengths**: *In the Euclidean space $\mathbb{R}^3$ every line with two endpoints has a finite length.*

Corollary 5.4 or Infinite Lengths: *In the Euclidean space $\mathbb{R}^3$ lines of infinite length are inconsistent.*

Theorem 4.2 of the Finite Divisions: *A finite interval of real numbers divided into parts of equal length, except at most the last part if any, can only have a finite number of parts.*

Theorem 4.3 of the Inconsistent divisions: *The Actual infinite division of any finite real interval is inconsistent.*

Theorem 5.8 of the Discrete Threshold: *The laws of physics do not apply in spaces smaller than the indivisible unit of space nor in times smaller than the indivisible unit of time, both being of non-zero extension (duration).*

Theorem 5.11 of Indivisible Units: *There is an indivisible minimum of space (time) intervals of which all space (time) intervals are an integer multiple.*

Theorem 5.6 of Finite Space and Time: *Every space interval (or time interval) is finite and can only be divided into an integer number of adjacent qusits (qutits), or into an integer number of adjacent parts, each an integer multiple of adjacent qusits (qutits).*

Theorem 8.1 of the Uncompletable Regress: *Any recursive sequence S of proofs, definitions or causes in which there is a last element to be proved (defined, caused) and each element has an immediate predecessor that proves (defines or causes) it, is uncompletable.*

Theorem 8.2 of the First Element: *A consistent sequence in which there is a last element and each element has an immediate predecessor is a complete totality only if it has a first element.*

16.4 The discrete solution of Kant's four antinomies

Making use now of the results that have just been recalled, it is immediate to prove the contradictory nature of Kant's four antinomies, pointing out, moreover, the cause of this contradictory nature. Indeed, consider the antithesis of Kant first antinomy:

> *Antithesis*: The world has no beginning and no limits
> in space, it is infinite, both in time and space.

According to Corollary 3.3, and considering the involved infinite is the actual infinity, this statement is false. So, the corresponding thesis

> *Thesis*: The world has a beginning in time and with
> respect to space it is also enclosed in limits.

could be true, which agrees with Theorem 3.12 and Theorem 5.1. The first Kant's antinomy is then a contradiction that results from the inconsistency of the Hypothesis of the Actual Infinity.

Let now consider the antithesis of the second antinomy:

> *Antithesis*: No compound thing in the world is made
> up of simple parts; and there is nothing simple in the
> world.

With respect to space and time, the above statement of the second antinomy clearly goes at least against Theorems 4.2, 4.3, Corollary 5.4 and Theorems 5.8, 5.11 and 5.6. With respect to any other physical object, and according to Corollary 3.3, Theorem 3.8 and Theorem 3.12, none of them can be consistently divided into an infinite number of parts. In addition, and now according to the Standard Model of Elementary Particles, all physical objects and any amount of mass and energy are made of a finite number of indivisible particles: quarks, leptons and bosons. Therefore, the antithesis of Kant's second antinomy is formally false, and its corresponding thesis:

> *Thesis*: Every compound substance in the world is
> composed of simple parts; and there is nothing but
> the simple or the compound of the simple

could be true, which agrees, on the one hand with all finitist results proved in this series of articles; and on the other with quantum physics and the Standard Model of elementary particles. In this case, it is again the inconsistency of the Hypothesis of the Actual Infinity that is the formal responsible for Kant's second contradiction.

With respect to Kant's third antinomy we must remember that in the universe there are natural informed systems that create, process and express physical information, being the expression of physical information the capacity to produce arbitrary properties: properties that, although they must be compatible with physical laws, cannot be deduced from physical laws but from the particular evolutionary history of each group of living beings. For example, the property of walking, or jumping, or singing, or of being self-conscious, which makes it possible to decide whether to do or not to do this or that, which makes freedom possible. Consequently, freedom is also a natural property of some natural objects as certain living beings. It is a consequence of the directional evolution of the universe that also makes possible the organic evolution of living beings in the direction of the increasing complexity [188, 189]. Consequently, the antithesis of Kant's third antinomy:

> *Antithesis*: There is no freedom whatsoever, but everything in the world happens only according to the laws of nature.

is also false. And the corresponding thesis:

> *Thesis*: Causality according to the laws of nature is not the only one from which the phenomena of the world can all be deduced. It is also necessary to admit, in order to explain them, a causality by freedom.

would be true by simply admitting the natural existence of informed living beings, capable of acquiring arbitrary properties, such as consciousness and freedom.

Kant's fourth antinomy is again related to the infinite, though now through the infinite regress of causes. According to the antithesis of this antinomy there would be no first cause of the world:

> *Antithesis*: There is nowhere an absolutely necessary being, either in the world or outside the world, as its cause.

But, according to the enormous inductive experience accumulated during the last two centuries, the universe evolves directionally towards maximum entropy (isotropy) and minimum temperature (Principle 5.1 of Directional Evolution). From this principle it immediately follows the Theorem 5.3 of the Consistent Universe, which in turn implies that no natural object can be the cause of itself (Theorem 5.6 of the Formal Dependence). The universe has a finite age and any instant of its history that we consider represents the last stage of its evolution until that instant. Therefore, being a consistent universe where no object originates itself, with a last stage (present) and each stage a consequence of a previous one, the complete history of the universe requires a first cause of its origin (Theorem 8.2 of the First Element). Which proves the falsity of the antithesis of Kan's fourth antinomy. Therefore, the corresponding thesis:

> *Thesis*: To the world belongs something which, as its part or as its cause, is an absolutely necessary being.

could be true.

Paper 17. A formal base for a discrete cosmology

Abstract.-This last paper of the series proposes the consideration of a set of formal elements to initiate a discussion on the foundations of a discrete cosmology. The main elements of this set will be the Principle of Directional Evolution and a short number of theorems and corollaries formally deduced from it. Taking into account the inconsistency of one of the fundamental axioms of the infinitist mathematical language of current cosmology (the Axiom of Infinity), the proposed consideration is formally justified. This paper 17, and with it this series of articles, ends with a list of the possible advantages of a finitist and discrete cosmology, compatible with all the empirical evidence about the observable universe, and oriented towards a more physical than mathematical interpretation of the great mysteries of contemporary physics.

Keywords: Theorem 5.5 of Identicality, Principle of Inertia, Principle of Consistency, Theorem of Formal Dependence, discrete paradigm, CALMs

17.1 An inductive principle for a discrete cosmology

Let us begin by recalling that in a finite and discrete universe there would be indivisible units of space and time. Although they have received different names, here we have proposed to call them respectively qusits (quantum space units) and qutits (quantum time units). Let us also remember that in the precedent articles of this series of articles an inductive principle of the maximum empirical evidence was formulated: the Principle (5.1) of Directional Evolution, which includes the independence from rational observers because for most of its history there were no rational observers in the universe:

The universe always evolves independently of its rational observers and in the same direction of increasing its global entropy.

This inevitable degradative evolution towards a final of maximum isotropic cold is compatible with the local and temporal appearance of ordered systems (minerals, for example) and of organized (teleological) systems, such as living beings. These are systems that maintain exchanges (physical and/or chemical) with their environment, and in such a way that the overall isotropic balance of both (system/environment) is always positive, in the direction indicated by the Principle of Directional Evolution. This fact should deserve the utmost consideration in any scientific cosmology. At the moment this is not the case.

17.2 A formal core for a discrete cosmology

As with other inductive principles, it is surprising the amount of information packed in the Principle of Directional Evolution. Information that can be extracted by successive applications of the basic laws of logic and basic logic inferences. Among (surely many) others, the following results follow almost immediately from both the Principle of Directional Evolution (also making use of the inconsistency of the actual infinity demonstrated in the paper 3 and in [198]):

- **Definition 3.10 of Discrete Set**: *A set is discrete if it has a first element, a last element and each of its elements (except the first one) has an immediate predecessor and (except the last one) an immediate successor.*

- **Theorem 3.9 of the Discrete Sets**: *All discrete sets are finite.*

- **Theorem 5.3 of the Consistent Universe**: *The universe evolves under the control of a unique set of invariant and consistent physical laws.*

- **Theorem 5.6 of the Formal Dependence**: *No concept defines itself; no statement proves itself; no physical object is the cause of itself; and no cause is the cause of itself.*

- **Theorem 5.8 of the Discrete Threshold**: *The laws of physics do not apply in spaces smaller than the minimum unit of space nor in times smaller than the minimum unit of time, both being of non-zero extension (duration).*

- **Corollary 5.5 of the Universal Physical Laws**: *The laws of* *physics apply to all regions of space and time, provided that they are greater than the minimum units of space and time.*

- **Theorem 5.11 of the Indivisible Units**: *All space regions and time intervals in which apply the laws of physics are integer multiples of their respective minimum units.*

- **Theorem 5.12 of Adjacency**: *No space exists between any two successive space minimum units, and no time elapses between two successive time minimum units.*

- **Corollary 5.6 of the Finite Space and Time**: *Every space interval (or time interval) is finite and can only be divided into an integer number of adjacent space (or time) minimum units.*

- **Theorem 6.1 of the Canonical Changes**: *Every change is either a canonical change of a discrete sequence of canonical changes.*

- **Theorem 6.2 of Change**: *Canonical changes are impossible in the spacetime continuum.*

- **Theorem 8.1 of the Uncompletable Regress**: *Any recursive sequence S of proofs, definitions or causes in which there is a last element to be proved (defined, caused) and each element has an immediate predecessor that proves (defines or causes) it, is uncompletable.*

- **Theorem 8.2 of the First Element**: *A consistent sequence in which there is a last element and each element has an immediate predecessor is a complete totality only if it has a first element.*

- **Theorem 9.1 of the Discrete Motion**: *The continuum densely ordered spacetime cannot be used to model uniform motion.*

- **subTheorem 9.1 of Physical Space and Time**: *The indivisible units of space and time are physical, and then real and absolute.*

- **Theorem 5.5 of Identicality:** *All particles of the same type have the same properties and behave the same way under the same conditions.*

- **Corollary 10.1 of Composed Identicality**: *The Theorem 5.5 of Identicality also applies to any combination of elementary particles, as atoms or molecules or even more complex particles, provided that the combinations are identical.*

- **Theorem 10.2 of the Arrow of Time**: *In a consistent universe the joint evolution of any system and its environment is always in the same direction of increasing its entropy.*

- **Theorem 10.3 of the Discrete History**: *The continuum densely ordered time cannot be used to model the history of physical objects.*

- **Theorem 11.1 of Abstract Points**: *In the spacetime continuum, the points of space have neither size nor shape, and the instants of time have not duration.*

Among others, the above formal results could be the initial building blocks for the design of a new finite and discrete model of the observable universe.

17.3 Preinertia: completing a mechanical principle

Preinertia was introduced in the article 7 of this series. It is a universal property of all physical objects, including photons. A property that has so far gone unnoticed by physics, although physics makes continuous (implicit, or best unconscious) use of it. And it is not an irrelevant property: It is nothing less than the reason why it is impossible to detect the absolute motion of a reference frame within that reference frame, i.e. with the only aid of the objects of that frame, including the photons produced in that frame (see the article 7 in this series of articles, and the corresponding chapters in [197]). Let us recall, then, that:

> *A physical object is said preinertial if it inherits the relative velocity vector of the reference frame where it is set in motion.*

In the aforementioned article 7 of this series of articles it was proved (making use of the speed of light as the universal constant c) that all physical objects are preinertial:

> **Theorem 7.1 of Preinertia**: *Every physical object inherits in one of its vector components the relative velocity vector of the reference system where it is set in motion, provided that the resulting speed does not exceed the possible maximum limit.*

On the other hand, there is the most overwhelming empirical evidence for preinertia, it is the reason why we fall exactly below the place where we jump vertically (don't forget that the Earth moves at 367 km/s in the galactic direction $(264.4, 48.4) \pm (0.3, 0.5)$ [118]). Such overwhelming empirical evidence and its predictable relation to inertia (yet to be established) point to the convenience of including Theorem 7.1 of Preinertia in the statement of the Principle of Inertia. An inclusion, on the other hand, very simple and straightforward:

> **Principle of Inertia**: *Every physical object **is preinertial and** remains at rest or moves at a constant uniform velocity, unless an external force acts upon it.*

17.4 CALM, a suggestion to start with

Naturally, everything remains to be done in the design of a new finite and discrete model for the observable universe. The above results only indicate a few things that should be taken into account in such a design. With the same guiding intention, in this series of articles we have considered the structure and operation of CALMs (cellular automata like models), essentially finite, discrete models that have been well known for several decades. Some of the advantages of this type of models could be the following:

1. Physics would no longer be dependent of an arbitrary mathematical axiom (the Axiom of Infinity), which could be inconsistent in accord with the Theorem 3.5 of the Denumerable Infinity (paper 3) and other arguments [198].

2. The still unresolved problem of change, perhaps the most basic problem in physics (although completely forgotten by physics and metaphysics), may finally be solved.

3. All the oddities of relativity could be only apparent, as the refractive deformation of a rod partially submerged in water [197]. Furthermore, the conversion factor between the geometry of a CALM and the geometry of the continuum is precisely the relativistic Lorentz factor.

4. Nothing can last a time less than one qutit nor move a distance less than one qusit. In a discrete model there is a maximum speed of one qusit per qutit. That could be the speed of light in the vacuum, though not necessarily.

5. The Second Principle of relativity would not be necessary because in a discrete space and time there is an insurmountable velocity of one qusit per qutit.

6. Unlike points and instants, the discrete units of space and time (qusits and qutits) are full of physical meaning.

7. CALMs are much more simple than the spacetime continuum: between any two points of the spacetime continuum infinitely

many other points exist, while the number of qusits in the whole visible universe would be finite ($\approx 7.64 \times 10^{184}$ if they were cubes of a Planck's volume).

8. Once replaced the spacetime continuum by a discrete space and time, quantum field theory (QFT) resembles the logic and functioning of a CALM.

9. Motion has not to be referred to abstract reference frames but to the actual fabric of space (although for practical reasons we could also make use of symbolic reference frames).

10. The entanglement of space and time in the spacetime continuum finds a natural explanation within the logic of CALMs functioning.

11. Quantum entanglement and quantum non-locality could be a natural consequence of CALMs synchronized way of functioning.

12. The flow of time and its irreversible arrow, enigmatic from the perspective of the spacetime continuum, is naturally explained in CALM terms.

13. The slippery concept of *now* could also be easily explained in CALM terms.

14. The incessant quantum activity of free space in QFT could be better explained in CALM terms than it is in the spacetime continuum.

15. Although QFT seems to be complete, the spacetime continuum it makes use of could be inappropriate. If that were the case, realist theories (and then the common sense in science) could recover its lost validity.

16. General relativity and QFT would have a new discrete opportunity to meet each other [9].

17. While points and instants of the spacetime continuum are primitive concepts devoid of physical meaning, and then hard to link with physical reality, qusits and qutits are plenty of physical significance.

18. All known physical objects and magnitudes, just except spacetime, are discrete, with indivisible units. In CALMs there is no exception, space and time are also discrete.

19. In the analog models of nature, extent and shape loss their physical meaning at the elementary particle level. This meaning could be found within the discrete space and time of CALMs.

20. Since each variable defining the state of a qusit is updated at each successive qutit, qusits can vibrate in multiple forms. And these vibrations could be the fundamental cause of the particle/wave duality in QFT.

21. Certain qusit states could have organizing effects and then could give rise to emergent objects and properties.

22. Observers, instruments and observed objects would form part, all of them, of the same CALM. Their current mutual interactions would determine their irreversible future.

But the most interesting consequences are probably not in the above list, but in the minds of some of its readers.

List of theorems

List of Figures

Bibliographic references

[1] A. D. Aczel. *The Mystery of the Aleph: Mathematics, the Kabbalah and the Search for Infinity.* Pockets Books, New York, 2000.

[2] A. Alemán. El realismo en matemáticas. *Mathesis,* 11(1):23 – 35, 1995.

[3] A. Alemán. El argumento de indispensabilidad en matemáticas. *Teorema,* XVIII(2):23–35, 1999.

[4] R. Aloisio, A. Galante, Grillo A., Luzio E., and F. Méndez. A note on DSR-like approach to space-time. *Phys. Lett. B,* 610:101 – 106, 2005.

[5] R. Aloisio, A. Galante, A. F. Grillo, E. Luzio, and F. Méndez. Approaching space-time through velocity in doubly special relativity. *Phys. Rev. D,* 70:125012, Dec 2004.

[6] J. S. Alper and M. Bridger. Mathematics, Models and Zeno's Paradoxes. *Synthese,* 110:143 – 166, 1997.

[7] J. S. Alper and M. Bridger. On the Dynamics of Perez Laraudogotia's Supertask. *Synthese,* 119:325 – 337, 1999.

[8] J. S. Alper, M. Bridger, J. Earman, and J. D. Norton. What is a Newtonian System? The Failure of Energy Conservation and Determinism in Supertasks. *Synthese,* 124:281 – 293, 2000.

[9] J. Ambjorn, J. Jurkiewicz, and R. Loll. El universo cuántico autoorganizado. *Investigación y Ciencia,* 384:20–27, 2008.

[10] Amelino-Camelia, Giovanni, Lämmerzahl, Claus, Mercati, Flavio, Tino, and Guglielmo M. Publisher's Note: Constraining the Energy-Momentum Dispersion Relation with

Planck-Scale Sensitivity Using Cold Atoms [phys. Rev. Lett. 103, 171302 (2009)]. *Phys. Rev. Lett.*, 104:039901, Jan 2010.

[11] G. Amelino-Camelia. A Phenomenological description of quantum gravity induced space-time noise. *Nature*, 410:1065–1069, 2001.

[12] C. Anastopoulos. *Particle or Wave. The evolution of the concept of matter in modern physics*. Princeton University Press, New Jersey, 2008.

[13] Aristotle. *Posterior Analytics*. Kessinger Publishing LLC, Whitefish, MT, 2004.

[14] Aristóteles. *Física*. Gredos (Kindle Edition), Madrid, 1995.

[15] Aristóteles. *Metafísica*. Espasa Calpe, Madrid, 1995.

[16] G. Arrigo and B. D'Amore. Lo veo pero no lo creo. Obstáculos epistemológicos y didácticos en el proceso de comprensión de un teorema de Cantor que involucra al infinito actual. *Educación matemática*, 11(1):5–24, 1999.

[17] J. Baez. The Quantum of Area? *Nature*, 421:702 – 703, February 2003.

[18] E. Baird. Two exact derivations of the mass/energy relationship, $E=mc^2$. *arXiv phisics physics/0009062*, pages 1–5, 2000.

[19] J. D. Barrow. *The Infinite Book*. Vintage Books (Random House), New York, 2006.

[20] Jacob D. Bekenstein. La información en un universo holográfico. *Investigación y Ciencia*, 325:36–43, 2003.

[21] Paul Benacerraf. Tasks, Super-Tasks, and Modern Eleatics. *J. Philos.*, LIX(24):765–784, 1962.

[22] Jean Paul Van Bendegem. In defense of discrete space and time. *Logique et Analyse*, 38:150 –152, 1997.

[23] Jean Paul Van Bendegem. Finitism in Geometry. In E. N. Zalta, editor, *Stanford Encyclopaedia of Philosophy*. Stanford University, URL = http://plato.stanford.edu, 2002.

[24] Bernadette Bensaude-Vincent. Lavoisier: una revolución científica. In Michel Serres, editor, *Historia de las ciencias*. Cátedra, Madrid, 1991.

[25] Henri Bergson. *Creative Evolution*. Dover Publications Inc., New York, 1998.

[26] Henri Bergson. The Cinematographic View of Becoming. In Wesley C. Salmon, editor, *Zeno's Paradoxes*, pages 59 – 66. Hackett Publishing Company, Inc, Indianapolis/Cambridge, 2001.

[27] George Berkeley. *A Treatise Concerning the Principles of Human Knowledge*. Renascence Editions, http://darkwing.uoregon.edu/ bear/berkeley, 2004.

[28] Alberto Bernabé. Introducción y notas. In Alberto Bernabé, editor, *Fragmentos presocráticos*. Alianza, Madrid, 1988.

[29] Károly Bezdek. *Classical Topics in Discrete Geometry*. Springer., New York, 2010.

[30] Joël Biard. Logique et physique de l'Infini au XIVe siècle. In Fran çoise Monnoyeur, editor, *Infini des mathématiciens, infinit des philosophes*. Belin, Paris, 1992.

[31] Erwin Biser. Discrete Real Space. *J. Philos.*, 38:518 – 524, 1941.

[32] Eftichios Bitsakis. Space and Time: Who was Right, Einstein or Kant? In Franco Selleri, editor, *Open Questions in Relativistic Physics*, pages 115–125. Apeiron, Montreal (Canada), 1998.

[33] M. Black. Achilles and the Tortoise. *Analysis*, XI:91 – 101, 1950 - 51.

[34] Simon Blackburn. *Oxford Dictionary of Philosophy*. Oxford University Press, New York, second edition edition, 2008.

[35] D. Blanco Laserna. *El bosón de Higgs. Los secretos de la partícula divina*. RBA Editores, Barcelona, 2015.

[36] D. Blanco Laserna. *Las ondas gravitacionales*. RBA Editores, Mexico, 2017.

[37] David Blanco Ledesma. *El espacio es una cuestión de tiempo*. RBA Editores, 2012.

[38] D. Bohm. *La totalidad y el orden implicado*. Editorial Kairós, 1987.

[39] Niels Bohr. *Atomic Physics and Human Knowledge*. John Wiley and Sons, New York, 1958.

[40] E. J. Borowski and J. M. Borwein. *The Harper Collins Dictionary of Mathematics*. Harper Collins Publisher, New York, 1991.

[41] Rodney A. Brooks. *Fields of Colors*. Rodeney A. Brooks, 2016.

[42] Antonopoulos. C. A Bang into Nowhere. Comments on the Universe Expansion Theory. *Apeiron*, 10(1):40–68, January 2003.

[43] A. Caba Sánchez. Algunas consideraciones sobre el argumento de indispensabilidad en matemáticas. *Revista de Filosofía*, 27(1):113–133, 2002.

[44] R. T. Cahill. Quantum Foam, Gravity and Gravitational Waves. *Relativity, Gravitation, Cosmology*, pages 168–226, 2003.

[45] Florian Cajori. The History of Zeno's Arguments on Motion. *American Mathematical Monthly*, XXII:1–6, 38–47, 77–82, 109–115, 143–149, 179,–186, 215–220, 253–258, 292–297, 1915. http://www.matedu.cinvestav.mx/librosydocelec/Cajori.pdf.

[46] Florian Cajori. The Purpose of Zeno's Arguments on Motion. *Isis*, III:7–20, 1920-1921.

[47] Georg Cantor. *Gesammelte Abhandlungen*. Verlag von Julius Springer, Berlin, 1932.

[48] Georg Cantor. *Contributions to the founding of the theory of transfinite numbers*. Dover, New York, 1955.

[49] Georg Cantor. Foundations of a General Theory of Manifolds. *The Theoretical Journal of the National Caucus of Labor Committees*, 9(1-2):69 – 96, January - February 1976.

[50] Georg Cantor. *Fundamentos para una teoría general de conjuntos*. Crítica, Barcelona, 2005.

[51] Alberto Casas González. *La materia oscura. El elemento más misterisos del universo*. RBA Editores, Barcelona, 2016.

[52] Ana María Cetto. *La luz. En la naturaleza y en el laboratorio*. Fondo de Cultura Económica, 2012.

[53] Li M. Chen. *Digital and discrete geometry*. Springer, New York, 2014.

[54] Marios Christodoulou and Carlo Rovelli. On the Possibility of Experimental Detection of the Discreteness of Time. *Frontiers in Physics*, 8:207, 2020.

[55] I. Ciufolini, V. Gorini, U. Moschella, and P. Fré. *Gravitationsl Waves*. Institute of Physics Publishing, 2001.

[56] Christopher Clapman and James Nicholson. *The Concise Oxforf Dictionary of Mathematics*. Oxford University Press, New York, 2009.

[57] Brian Clegg. *A Brief History of Infinity. The Quest to Think the Unthinkable*. Constable and Robinson Ltd, London, 2003.

[58] Frank Close. *Particle Physics. A Very Short Introduction*. Oxford University Press, New York, 2004.

[59] Frank Close. *Nothing*. Oxford University Press, New York, 2009.

[60] Jonas Cohn. *Histoire de l'infini. Le problème de l'infini dans la pensée occidentale jusqu''a Kant*. Leséditions du CERF, Paris, 1994.

[61] H. R. Coish. Elementary particles in a finite world geometry. *Phys. Rev.*, 114:383 – 388, 1959.

[62] Giorgio Colli. *Zenón de Elea*. Sexto Piso, Madrid, 2006.

[63] Lucas C. Céleri and Vasileios I. Kiosses. Unruh effect as a result of quantization of spacetime. *Physics Letters B*, 781:611 – 615, 2018.

[64] John Daintith, editor. *Dictionary of Physics*. Oxford University Press, New York, 2009.

[65] Josep W. Dauben. *Georg Cantor. His mathematics and Philosophy of the Infinite*. Princeton University Press, Princeton, N. J., 1990.

[66] San Anselmo de Canterbury. *Proslogion. Con las réplicas de Gaunilón y Anselmo*. Ed. Tecnos, 2009.

[67] Agustín de Hipona. *Confesiones*. Editorial Gredos, Madrid, 2010 (397-398).

[68] Richard Dedekind. *Qué son y para qué sirven los números (Was sind Und was sollen die Zahlen (1888))*. Alianza, Madrid, 1998. Definición de conjunto infinito p. 115.

[69] Jean-Paul Delahaye. El carácter paradójico del Infinito. *Investigación y Ciencia (Scientifc American)*, Temas: Ideas del infinito(23):36 – 44, 2001.

[70] Satyan L. Devadoss and Joseph O'Rourke. *Discrete and computational geometry.* Princeton University Press, Princeton and Oxford, 2011.

[71] P. A. M. Dirac. *Quantum Mechanics.* Oxford University Press, London, 1958.

[72] Charles L. Dogson. *Euclid and His Modern Rivals.* Forgotten Books, London, 2013.

[73] John Earman. Determinism: What We Have Learned and What We Still Don't Know. In Michael O'Rourke and David Shier, editors, *Freedom and Determinism,* pages 21–46. MIT Press, Cambridge, 2004.

[74] John Earman and John D. Norton. Forever is a Day: Supertasks in Pitowsky and Malament-Hogarth Spacetimes. *Philosophy of Science,* 60(1):22–42, 1993.

[75] John Earman and John D. Norton. Infinite Pains: The Trouble with Supertasks. In S. Stich, editor, *Paul Benacerraf: The Philosopher and His Critics.* Blackwell, New York, 1996.

[76] John Earman and John D. Norton. Comments on Laraudogoitia's 'Classical Particle Dynamics, Indeterminism and a Supertask'. *The British Journal for the Phylosophy of Science,* 49(1):122 – 133, March 1998.

[77] A. Einstein. Zur Elektrodynamik bewegter Körper. *Ann. Phys.,* 17:891–921, 1905.

[78] A. Einstein. Über den Einfluss der Schwercraft auf die Ausbreitung des Lichtes. *Ann. Phys.,* 35:898–90, 1911.

[79] A. Einstein. Zum Relativitätsproblem. *Scientia,* 15:337–348, 1914.

[80] A. Einstein. Die Grundlage der allgemeinen Relativitätstheorie. *Ann. Phys.,* 354(7):769–822, 1916.

[81] A. Einstein. *The Meaning of Relativity. Four Lectures Deliverd at Princeton University, May 1921.* Princeton University Press, 1922.

[82] A. Einstein. Cosmological considerations on the general theory of relativity. In *The Principle of Relativity.* Dover Publications Inc., 1952.

[83] A. Einstein. *The Principle of Relativity,* chapter III On the Electrodynamics of Moving Objects, pages 35–65. Dover Publications Inc., 1952.

[84] A. Einstein. The Problem of Space, Ether, and the Field In Physics. In Nick Huggett, editor, *Space from Zeno to Einstein*, pages 253 – 260. The MIT Press, Cambridge, 2002.

[85] A. Einstein. Sobre la electrodinámica de los cuerpos en movimiento. In A. Ruíz de Elvira, editor, *Cien años de relatividad. Los artículos clave de 1905 y 1906*, pages 88–139. nivola, Madrid, 2003.

[86] A. Einstein and L. Infeld. *The evolution of physics*. Cambridge University Press, 1938.

[87] A. Einstein and L. Infeld. *La evolución de la física*. Salvat Editores, Barcelona, 1986.

[88] A. Einstein, B. Podolsky, and N. Rosen. Can quantum mechanical description of physical reality be considered complete? *Physical Review*, 47:777– 780, May 1935.

[89] V. Ekroll, F. Faul, and J. Golz. Classification of apparent motion percepts basedon temporal factors. *Journal of Vision*, 8(4):1–22, 2008.

[90] Walter M. Elsasser. *Átomo y organismo. Nuevo enfoque de la Biología Evolucionista*. Siglo XXI, México, 1969.

[91] C. Emmeche, S. Koppe, and F. Stjernfelt. Explaining Emergence. *Journal for General Philosophy of Science*, 28:83–119, 1997.

[92] Berthold-Georg Englert, Marlan O. Scully, and Artur K. Ekert. La dualidad en la materia y en la luz. In *Misterios de la física cuántica*, pages 68–74. Prensa Científica. Investigación y Ciencia. Temas 10, 1997.

[93] Real Academia Española. *Diccionario de la lengua española*. Real Academia Española, 2014.

[94] Euclid. *Elements*. Green Lion Press, 2002.

[95] Euclides. *Elementos*. Gredos, Madrid, 2000.

[96] Michael Faraday. On the Physical Character of the Lines of Magnetic Force. *The London, Edinburg and Doublin Philosophical Magazine and Journal of Science*, 3(20):407–437, 1852.

[97] Michael Faraday. *On the Physical Character of the Lines of Magnetic Firce*. HardPress, Miami, 2017.

[98] A. A. Faraj. The Ives Experiment. *The General Science Journal*, pages 1–15, 2016.

[99] Jesús Navarro Faus. *Heisenberg. El principio de incertidumbre*. RBA Editores, Barcelona, 2012.

[100] José L. Fernández Barbón. Geometría no conmutativa y espaciotiempo cuántico. *Investigación y Ciencia*, (342):60–69, Marzo 2005.

[101] Enrique Fernández Borja. *El vacío y la nada*. RBA Editores, Barcelona, 2015.

[102] Ramón Fernández Álvarez Estrada and Marina Ramón Medrano. *Partículas elementales*. Editorial Pirámide, 2018.

[103] A. Ferrer Soria and Eduardo Ros Martínez. *Física de partículas y de astropartículas*. Universitat de Valencia, 2014.

[104] Richard P. Feynman. *Electrodinámica cuántica: la extraña teoría de la luz y la materia*. Alianza Universidad, Madrid, 1992.

[105] Richard P. Feynman. *El carácter de la ley física*. Tusquets, Barcelona, 2000.

[106] Richard P. Feynman, Robert B. Leighton, and Matthew Sands. *Lectures on Physics*. Addison-Wesley Publishing Company, Reading, Massachusetts, 1977.

[107] R.P. Feynman and S. Weinberg. *LAS PARTÍCULAS ELEMENTALES Y LAS LEYES DE LA FÍSICA*. Gedisa, Barcelons, 1997.

[108] D. Finkelstein and E. Rodríguez. Quantum time-space and gravity. In R. Penrose and C. J. Isham, editors, *Quantum Concepts in Space and Time*, pages 247 – 254. Oxford University Press, Oxford, 1986.

[109] Fabrizio Fiore, Luciano Burderi, Tiziana Di Salvo, Marco Feroci, Claudio Labanti, Michelle R. Lavagna, and Simone Pirrotta. HERMES: a swarm of nano-satellites for high energy astrophysics and fundamental physics. In Jan-Willem A. den Herder, Shouleh Nikzad, and Kazuhiro Nakazawa, editors, *Space Telescopes and Instrumentation 2018: Ultraviolet to Gamma Ray*, volume 10699. International Society for Optics and Photonics, SPIE, 2018.

[110] E. Fischbach, H. Kloor, R. A. Langel, A. T. Y. Lui, and M. Peredo. New geomagnetic limits on the photon mass and on long-range forces coexisting with electromagnetism. *Phys. Rev. Lett.*, 73:514–517, Jul 1994.

[111] Robert Fogelin. Hume and Berkeley on the Proofs of Infinite Divisibility. *The Philosophical Review*, XCVII(1):47 – 69, January 1988.

[112] P. Forrest. Is Space-Time Discrete or Continuous? *Synthese*, 103:327 – 354, 1995.

[113] A. Friedmann. Über die Krümmung des Raumes. *Zeitschrift für Physik*, 10 (1):377–386, June 1922.

[114] A. Friedmann. Über die Möglichkeit einer Welt mit konstanter negativer Krümmung des Raumes. *Zeitschrift für Physik*, 21 (1):326–332, 1924.

[115] G. Galilei. *Dialogue concerning the two chief world systems-Ptolemaic and Copernican*. University of California Press, Berkeley and Los Angeles, 1967.

[116] C. F. Gauss and H. C. Schumacher. *C. F. Gauss Werke: Briefwechsel Mit Schumacher 3 vol.* Georg Olms, 1831 (1075).

[117] Henning Genz. *Nothingness. The science of empty space.* Perseus Publishing, Cambridge, MA, 1999.

[118] Cameron Gibelyou and Dragan Huterer. Dipoles in the sky. *Monthly Notices of the Royal Astronomical Society*, 427(3):1994–2021, Nov 2012.

[119] P. Glansdorff and I. Prigogine. *Structure, Stabilité et Fluctuations.* Mason, Paris, 1971.

[120] James Gleick. *The information.* Fourth Estate, London, 2012.

[121] Pedro F. González Díaz. La energía fantasma y el futuro del universo. *Investigación y Ciencia*, (357):53–61, Junio 2006.

[122] I. Grattan-Guiness. *Del cálculo a la teoría de conjuntos, 1630-1910.* Alianza Editorial, Madrid, 1984.

[123] Marvi Jay Greenberg. *Euclidean and Non-Euclidean Geometries. Development and History.* W.H. Freeman, 2008.

[124] Brian Greene. *The Fabric of the Cosmos. Space, Time, and the Texture of Reality.* Alfred A. Knopf, New York, 2004.

[125] Adolf Grünbaum. Modern Science and Refutation of the Paradoxes of Zeno. *The Scientific Monthly*, LXXXI:234–239, 1955.

[126] Adolf Grünbaum. *Modern Science and Zeno's Paradoxes*. George Allen And Unwin Ltd, London, 1967.

[127] Adolf Grünbaum. Modern Science and Refutation of the Paradoxes of Zeno. In Wesley C. Salmon, editor, *Zeno's Paradoxes*, pages 164 – 175. Hackett Publishing Company, Inc, Indianapolis/Cambridge, 2001.

[128] Adolf Grünbaum. Modern Science and Zeno's Paradoxes of Motion. In Wesley C. Salmon, editor, *Zeno's Paradoxes*, pages 200 – 250. Hackett Publishing Company, Inc, Indianapolis/Cambridge, 2001.

[129] Adolf Grünbaum. Zeno's Metrical Paradox of Extension. In Wesley C. Salmon, editor, *Zeno's Paradoxes*, pages 176 – 199. Hackett Publishing Company, Inc, Indianapolis/Cambridge, 2001.

[130] Alan Guth. *The Inflationary Universe. The Quest for a New Theory of Cosmics Origins*. Basic Books, 1998.

[131] Alan H. Guth. Inflationary universe: A possible solution to the horizon and flatness problems. *Phys. Rev. D*, 23:347–356, Jan 1981.

[132] Shahen Hacyan. *Física y metafísica del espacio y el tiempo*. Fondo de Cultura Económica, Mexico, 2004.

[133] Michael Hallet. *Cantorian Set Theory and Limitation of Size*. Oxford University Press, 1984.

[134] Charles Hamblin. Starting and Stopping. *The Monist*, 53:410 –425, 1969.

[135] Stephen Hawking and Leónard Mlodinov. *El gran diseño*. Editorial Crítica S.L., Barelona, 2010.

[136] Stephen W. Hawking. *Historia del tiempo. Del big bang a os agujeros negros*. Crítica, Baercelona, 1988.

[137] Stephen W. Hawking. *El futuro del espaciotiempo*. Crítica, Barcelona, 2003.

[138] T. Heath. *The thirteen books of Euclid's Elements*, volume II. Cambridge University Press, Cambridge, second edition, 1926.

[139] Thomas Heath. *The Thirteen Books of Euclid's Elements*, volume I. Dover Publications Inc, New York, second edition, 1956.

[140] Thomas Heath. *A History of Greek Mathematics*. Dover Publications Inc, New York, 1981.

[141] Georg W.F. Hegel. *The Science of Logic*. Cambridge University Press, Cambridge, UK, 2010.

[142] Georg Wilhelm Frederich Hegel. *Lógica*. Folio, Barcelona, 2003.

[143] Michael Heller. *On Space and Time*, chapter Where physics meets metaphysics, pages 238–277. Cambridge University Press, 2008.

[144] David Hilbert. *The Foundations of Geometry*. The Open Court Publishing Company, La Salle, Illinois, 1950.

[145] Craig J. Hogan. *El libro del Big Bang*. Alianza, Madrid, 2005.

[146] Craig J. Hogan. Interferometers as Probes of Planckian Quantum Geometry. *Phys.Rev.*, D85:064007, 2012.

[147] M. L. Hogarth. Does General Relativity Allow an Observer to view an Eternity in a Finite Time? *Foundations of Physics Letters*, 5:173 – 181, 1992.

[148] Gerard't Hooft. *Partículas elementales*. Crítica, Barcelona, 1991.

[149] John Horgan. Filosofía cuántica. In *Misterios de la física cuántica*, pages 36–45. Prensa Científica. Investigación y Ciencia. Temas 10, 1997.

[150] S. Hossenfelder. Interpretation of quantum field theories with a minimal length scale. *Phys. Rev. D*, 73:105013, May 2006.

[151] Fred Hoyle. *El universo: galaxias, núcleos y quásares*. Alianza Editorial, 1967.

[152] Pamela H. Huby. Kant or Cantor? That the universe, if real, must be finite In both space and time. In A. W. Moore, editor, *Infinity*, pages 121 –152. Dartmouth, Aldershot, 1993.

[153] Nick Huggett. *Space from Zeno to Einstein*. MIT Press, Cambridge, Massachusetts, 2002.

[154] Nick Huggett. Zeno's Paradoxes. In Edward N. Zalta (ed.), editor, *The Stanford Encyclopaedia of Philosophy (Summer 2004 Edition)*. Stanford University, 2004.

[155] David Hume. *A Treatise of Human Nature*. Outlook Verlag GmbH, Frankfurt am Main, Germany, 2020.

[156] H. E. Ives and G. R. Stilwell. An experimental study of the rate of a moving atomic clock. II. *Journal of the Optical Society of America*, 31(5):369, 1941.

[157] Herbert E. Ives. Light Signals Sent Around a Closed Path. *Journal of the Optical Society of America*, 28(8):296–299, Aug 1938.

[158] Herbert E. Ives. Xlviii. derivation of the lorentz transformations. *The London, Edinburgh, and Dublin Philosophical Magazine and Journal of Science*, 36(257):392–403, 1945.

[159] Herbert E. Ives. Historical Note on the Rate of a Moving Atomic Clock. *Journal of the Optical Society of America*, 37(10):810–813, Oct 1947.

[160] Herbert E. Ives. The Measurement of the Velocity of Light by Signals Sent in One Direction. *Journal of the Optical Society of America*, 38(10):879–884, Oct 1948.

[161] Herbert E. Ives. Lorentz-Type Transformations as Derived from Performable Rod and Clock Operations. *Journal of the Optical Society of America*, 39(9):757–761, 1949.

[162] Herbert E. Ives. Revisions of the Lorentz transformation. *Proceedings of the American Philosophical Society*, 95(2):125–131, 1951.

[163] Herbert E. Ives and G. R. Stilwell. An Experimental Study of the Rate of a Moving Atomic Clock. *Journal of the Optical Society of America*, 28(7):215–226, Jul 1938.

[164] N. Jafari and A. Shariati. Projective interpretation of some doubly special relativity theories. *Phys. Rev. D*, 84:065038, Sep 2011.

[165] Max Jammer. *Concepts of Space: The History of Theories of Space in Physics*. Dover Publlcations, Inc., New York, 1993.

[166] Max Jammer. *Concepts of Mass in Contemporary Physics and Philosophy*. Princeton University Press, Princeton, New Jersey, 2000.

[167] A. Jannussis. Einstein and the Development of Physics. In Franco Selleri, editor, *Open Questions on Relativistic Physics*, pages 127–130. Apeiron, Montreal (Canada), 1998.

[168] I. Kant. *Crítica de la razón pura*. Editorial Hausben, 2011.

[169] Immanuel Kant. *Crítica del juicio*. Espasa Calpe, Madrid, 1984.

[170] Immanuel Kant. *Crítica de la razón pura.* Alfaguara, Madrid, 1989.

[171] Immanuel Kant. *Critique of pure reason.* Cambridge University Press, 1998.

[172] Hugh Kearney. *Orígenes de la ciencia moderna: 1500 - 1700.* Guadarrama, Madrid, 1970.

[173] Daniel Kennefick. Einstein Versus the Physical Review. *Physics Today,* 58(9):43, 2005.

[174] Ludwik Kostro. The Physical and Philosophical Reasons for A. Einstein Denial of the Ether in 1905 and its Reintroduction in 1916. In Franco Selleri, editor, *Open Questions in Relativistic Physics,* pages 131–139. Apeiron, Montreal (Canada), 1998.

[175] Ludwik Kostro. *Einstein and the ether.* Apeiron, Montreal, 2000.

[176] H. Kragh and B. Carazza. From Time Atoms to Space-Time Quantization: the Idea of Discrete Time, ca 1925-1936. *Studies in History and Philosphy of Science,* 25:437 – 462, 1994.

[177] Lawrence M. Krauss. *Un universo de la nada.* Pasado y Presente, S.L, Barcelona, 2013.

[178] Manuel Lallena Rojo. *El Big Bang y el origen del universo.* RBA Editores, Barcelona, 2015.

[179] Robert Laughlin. *Un universo diferente. La reinvención de la física en la edad de la emergencia.* Katz, Buenos Aires, 2007.

[180] Shaughan Lavine. *Understanding the Infinite.* Harvard University Press, Cambridge MA, 1998.

[181] Marc Lchièze-Rey. *L' infini. De la philosophie à l'astrophysique.* Hatier, Paris, 1999.

[182] León Lederman and Dick Teresi. *LA PARTÍCULA DIVINA Si el universo es la pregunta ¿cuál es la respuesta?* Editorial Crítica, Barcelona, 2008.

[183] G. Lemaître. Un univers homogène de masse constante et de rayon croissant, randant compte de la vitesse radiale des nèbuleuses extra-galactiques. *Annales de la Société Scientifique de Bruxelles,* 47:49–59, 1927.

[184] G. Lemaître. L'expansion de l'espace. *Publications du Laboratoire d'Astronomie et de Géodésie de l'Université de Louvain*, 8:101–120, 1931.

[185] G. Lemaître. L'univers en expansion. *Annales de la société scientifique de Bruxelles*, 53A:51–83, 1933.

[186] G. Lemaître. Expansion de l'Univers. *L'Astronomie*, pages 153–168, 1935.

[187] G. Lemaître. *L'Hypothèse de l'atome primitif, essai de cosmogonie*. Dunod, 1946.

[188] A. León Sánchez. Coevolution: New Thermodynamic Theorems. *J. Theor. Biol.*, 147(2):205 – 212, 1990.

[189] A. León Sánchez. Living beings as informed systems: towards a physical theory of information. *Journal of Biological Systems*, 4(4):565 – 584, 1996.

[190] A. León Sánchez. The aleph-zero or zero dichotomy. *Cogprints*, pages 1–7, September 2006. https://arxiv.org/abs/0804.2934.

[191] A. León Sánchez. Infinity one by one. *Preprint*, 2021.

[192] A. León Sánchez. A critique of selfreference: what Gödel theorem really proves. *The General Science Journal*, pages 1–9, 2021. PDF.

[193] A. León Sánchez. *New Elements of Euclidean Geometry*. Amazon's Kindle Direct Publishing, 2021. PDF.

[194] A. León Sánchez. *Paradoxes and theorems*. Amazon's Kindle Direct Publishing, 2021. PDF.

[195] A. León Sánchez. *The Physical Meaning of Entropy*. Amazon's KDP, 2021. PDF.

[196] A. León Sánchez. Contributions to the history, logic and philosophy of science 1/10. On the physical notions of order and organization. *The General Science Journal*, 2022. PDF.

[197] A. León Sánchez. *Apparent relativity*. Amazon's KDP, 2022. PDF.

[198] A. León Sánchez. *Infinity put to the test*. Amazon's KDP, 2023 (2021). PDF.

[199] A. León Sánchez. The shame of physics. *The General Science Journal*, 2023. PDF.

[200] Keren Li, Youning Li, Muxin Han, Sirui Lu, Jie Zhou, Dong Ruan, Guilu Long, Yidun Wan, Dawei Lu, Bei Zeng, and Raymond Laflamme. Quantum spacetime on a quantum simulator. *Communications Physics*, 2(1):122, 2019.

[201] O. J. Lodge. *The Ether of Space*. Alpha Editions, 2021.

[202] John Losee. *Introducción histórica a la filosofía de la ciencia*. Alianza, Madrid, 1987.

[203] Seth Loyd and Y. Jack Ng. Computación en agujeros negros. *Investigación y Ciencia (Scientifc American)*, (340):59 – 67, Enero 2005.

[204] Juan de Lugo. *Cómo se puede explicar la composición del continuo por solo indivisibles finitos, según la opinión de los filósofos actuaes*, chapter Las categorías de tiempo y espacio en el pensamiento de la Escolástica tardía, pages 84–106. Ediciones Universidad de Salamanca, Salamanca, España, 2004.

[205] Jean Pierre Lumient and Marc Lachièze-Rey. *La physique et l'infini*. Flammarion, Paris, 1994.

[206] Jean-Pierre Luminet, Glenn D. Starkman, and Jeffrey R. Weeks. ¿Es finito el espacio? *Investigación y Ciencia*, (273):6–13, Junio 1999.

[207] Peter Lynds. Time and Classical and Quantum Mechanics: Indeterminacy vs. Discontinuity. *Foundations of Physics Letters*, 16:343 – 355, 2003.

[208] Peter Lynds. Zeno's Paradoxes: A Timely Solution. *philsci-archives*, pages 1 – 9, 3003. http://philsci-archives.pitt.edu/archive/00001197.

[209] J. M. López Piñero, V. Navarro, and E. Portela. *La Revolución Científica*. Historia 16, Madrid, 1989.

[210] Qing-Ping Ma. Is E=mc^2 an exclusively relativistic result? *arXiv*, pages 1–20, 2015.

[211] Shahn Majid. Quantum space time and physical reality. In Shahn Majid, editor, *On Space and Time*, pages 56–140. Cambridge University Press, New York, 2008.

[212] Eli Maor. *To Infinity and Beyond. A Cultural History of the Infinite*. Pinceton University Press, Princeton, New Jersey, 1991.

[213] Mariano Martínez. *Espacio y tiempo en la Escuela de Salamanca*, chapter Prólogo, pages 11 – 16. Ediciones Universidad de Salamanca, Salamanca, 2004.

[214] Joel Gabàs Masip. *Maxwell. La naturaleza de la luz*. Nivola, 2012.

[215] Stephen F. Mason. *Historia de las ciencias*. Alianza, Madrid, 1985.

[216] Tim Maudlin. *Filosofía de la física I. El espacio y el tiempo*. Fondo de Cultura Económica, México, 2014.

[217] Tim Maudlin. *Philosophy of Pysics. Space and Time*. Princeton University Press, New Jersey, 2015.

[218] James Clerk Maxwell. *Materia y movimiento*. Crítica, Barcelona, 2006.

[219] Ernst Mayr. *This is Biology. The Science of Living World*. Harvard University Press, Cambridge, 1998.

[220] Joseph Mazur. *The Motion Paradox*. Dutton, 2007.

[221] William I. McLaughlin. Una resolución de las paradojas de Zenón. *Investigación y Ciencia (Scientifc American)*, (220):62 – 68, Enero 1995.

[222] William I. McLaughlin and Silvia L. Miller. An Epistemological Use of non-Standard Analysis to Answer Zeno's Objections Against Motion. *Synthese*, 92(3):371 – 384, September 1992.

[223] J. E. McTaggart. The unreality of time. *Mind*, 17:457 – 474, 1908.

[224] Brian Medlin. The Origin of Motion. *Mind*, 72:155 – 175, 1963.

[225] A. Meessen. Is it logically possible to generalize physics through space-time quantization? In P. Weingartner and G. Schurz, editors, *Philosophie der Naturwissenschaften. Akten des 13 Internationalen Wittgenstein Symposium*, pages 19 – 47. Hölder-Pichler-Tempsky, Vienna, 1989.

[226] Constantin Meis. *Light and Vacuum. The Wave-Particle Nature of the Light and the Quantum Vacuum*. World Scientific Publishing, 2017.

[227] N. David Mermin. *It's about time. Understanding Einstein's relativty*. Princeton University Press, Princetona and Oxford, 2009.

[228] C. W. Misner, K. S. Thorne, and J. A. Wheeler. *Gravitation*. W. H. Freeman and Company, San Francisco, 1973.

[229] Jacques Monod. *Chance and Necessity*. Alfred A. Knopf Inc., New York, 1971.

[230] Andreas W. Moore. Breve historia del infinito. *Investigación y Ciencia (Scientifc American)*, (225):54 – 59, 1995.

[231] Andreas W. Moore. *The Infinite*. Routledge, New York, 2001.

[232] Richard Morris. *Achilles in the Quantum Universe*. Henry Holt and Company, New York, 1997.

[233] Richard Morris. *La historia definitiva del infinito*. Ediciones B S.A., 2000.

[234] Chris Mortensen. Change. In E. N. Zalta, editor, *Stanford Encyclopaedia of Philosophy*. Stanford University, URL = http://plato.stanford.edu, 2020.

[235] Morveau, Bertholet, Furcroy, and Lavoisier. *Método de nomenclatura química*. Don Antonio de Sancha, 1788.

[236] Michael Moyer. Is space digital? *Sci. Amer.*, 306(2):30–37, 2012.

[237] Robert Munafo. Notable Properties of Specific Numbers, 2013.

[238] George Musser. Filosofía del tiempo. *Investigación y Ciencia (Scientifc American)*, (314):14 – 15, Noviembre 2002.

[239] Isaac Newton. *Mathematical Principles of Natural Philosophy*. Daniel Adee Publishing, New York, 1846.

[240] Isaac Newton. *Principios matemáticos de la filosofía natural*. Alianza, Madrid, 1987.

[241] John D. Norton. A Quantum Mechanical Supertask. *Found. Phys.*, 29(8):1265 – 1302, 1999.

[242] J.J. O'Connor and E.F. Robertson. Non-Euclidean Geometry History, 1996. McTutor History of Mathematics. Accesed: 2016-02-16.

[243] Javier Ordoñez, Victor Navarro, and José Manuel Sánchez Ron. *Historia de la Ciencia*. Espasa Calpe, Madrid, 2004.

[244] W.T. Padgett. Problems with the Current Definition of Mass. *Physics Essays*, 3:178–182, 1990.

[245] A. Pais. *Subted is the Lord. The Science and Life of Albert Einstein*. Oxford University Press, New York, 1982.

[246] Alba Papa-Grimaldi. Why mathematical solutions of Zeno's paradoxes miss the point: Zeno's one and many relation and Parmenides prohibition. *The Revew of Metaphysics*, 50:299–314, December 1996.

[247] Parménides. Acerca de la naturaleza. In Alberto Bernabé, editor, *De Tales a Demócrito. Fragmentos presocráticos*, pages 159 – 167. Alianza, Madrid, 1988.

[248] H. H. Pattee. El problema de la jerarquía biológica. In C. H. Waddington, editor, *Hacia una Biología Teórica*. Alianza, Madrid, 1970.

[249] H. H. Pattee. Las bases físicas de la codificación y fidelidad en la evolución biológica. In C. H. Waddington, editor, *Hacia una Biología Teórica*, pages 81–114. Alianza, Madrid, 1979.

[250] H. H. Pattee. Evolving self-reference: matter, symbols and semantic closure. *Comunication and Cognition*, 12:9–27, 1995.

[251] H. H. Pattee. The Problem of Observables in Models of Biological Organizations. In E. L. Khalil and E. Boulding, editors, *Evolution, Order and Complexity*. Routledge, London, 1996.

[252] Giuseppe Peano. *Arithmetices Principia. Nova Methodo Exposita*. Libreria Bocca, Roma, 1889.

[253] Alejandro Perez and Salvatore Ribisi. Energy-mass equivalence from Maxwell equations. *arXiv*, pages 1–17, 2021.

[254] Igor Pikovski, Michael R. Vanner, Markus Aspelmeyer, M. S. Kim, and Caslav Brukner. Probing Planck-scale physics with quantum optics. *Nat. Phys.*, 8:393–397, 2012.

[255] I. Pitowsky. The Physical Church Thesis and Physical Computational Complexity. *Iyyun: The Jerusalem Philosophical Quarterly*, 39:81 –99, 1990.

[256] Max Planck. Zur Theorie des Gesetzes der Energieverteilung im Normalspektrum. *Verh. Dtsch. Phys. Ges.*, 2:237–243, 1900.

[257] J. Playfair. *Elements of Geometry*. W.E. Dean Printer and Publisher, New York, 1846.

[258] John Playfair. *Elements of Geometry*. Number London. Forgotten Books, 2015.

[259] M. Polanyi. Life's Irreducible Structure. *Science*, 160:1308, 1968.

[260] Mary Potter, Carl Hagmann, and Emily McCourt. Banana or fruit? Detection and recognition across categorical levels in RSVP. *Psychonomic Bulletin and Review*, 22(2):578–585, 2015.

[261] H. Putnam. *Philosophy of Logic*. George Allen and Unwin, London, 1972.

[262] Jon Pérez Laraudogoitia. A Beautiful Supertask. *Mind*, 105:49–54, 1996.

[263] Jon Pérez Laraudogoitia. Classical Particle Dynamics, Indeterminism and a Supertask. *British Journal for the Philosophy of Science*, 48(1):49 – 54, 1997.

[264] Jon Pérez Laraudogoitia. Infinity Machines and Creation Ex Nihilo. *Synthese*, 115:259 – 265, 1998.

[265] Jon Pérez Laraudogoitia. Why Dynamical Self-Excitation is Possible. *Synthese*, 119(3):313 – 323, 1999.

[266] Jon Pérez Laraudogoitia. Supertasks. In E. N. Zaltax, editor, *The Stanford Encyclopaedia of Philosophy*. Standford University, URL = http://plato.stanford.edu, 2001.

[267] Jon Pérez Laraudogoitia, Mark Bridger, and Joseph S. Alper. Two Ways of Looking at a Newtonian Supertask. *Synthese*, 131(2):157 – 171, 2002.

[268] Martin Rees. *Just Six Numbers. The deep forces that shape the universe*. Phoenix. Orion Books Ltd., London, 2000.

[269] Hans Reichenbach. *The Philosophy of Space and Time*. Dover Publications Inc, New York, 1957.

[270] Thomas Reid. *Inquiry Into the Human Mind: on the Principle of Common Sense*. Edinburgh University Press, 1997 (1764).

[271] Jesus Rodríguez-Quintero. *MÁS ALLÁ DEL BOSON DE HIGGS. En busca de nuevas partículas*. RBA Editores México, 2019.

[272] José Rodríguez-Quintero. *Quarks y gluones. Las entrañas de las partículas elementales.* RBA Editores, Barcelona, 2016.

[273] F.J. Romero Mora and García García J.A. *La física de la luz.* RBA Editores, Barceloma, 2017.

[274] Michele Ronco. *Quantum Space-Time: theory and phenomenology.* PhD thesis, Rome U., 2018.

[275] Boris Abramovich Rosenfeld. *A History of Non-Euclidian Geometry. Evolution of the Concept of a Geometric Space.* Spriger Verlag, New York, 1988.

[276] Francesc Rossell i Pujols. *El infinito.* EMSE EDAPP, Barcelona, 2019.

[277] Tony Rothman. The Secret History of Gravitational Waves. *American Scientist*, 106(2):95, 2018.

[278] Brian Rotman. *The Ghost in Turing Machine.* Stanford University Press, Stanford, 1993.

[279] Carlo Rovelli. Quantum spacetime: What do we know? In Craig Callender and Nick Huggett, editors, *Physics meets Philosophy at the Plank scale*, pages 101 – 122. Cambridge University Press, Cambridge, 2001.

[280] Carlo Rovelli. *La realidad no es lo que parece. La estructura elemental de las cosas.* Tusquets, 2015.

[281] Chandrasekhar Roychoudhuri, A.F. KrackLauer, and Hatherine Creath, editors. *The nature of light. What is a photon?* CRC Press, 2019.

[282] José Alberto Rubiño. *El fondo cósmico de microondas.* RBA Editores, Barcelona, 2017.

[283] Rudy Rucker. *Infinity and the Mind.* Princeton University Press, Princeton, 1995.

[284] Bertrand Russell. *Misticismo y lógica y otros ensayos.* Aguilar, Madrid, 1973.

[285] Bertrand Russell. Sobre la inducción. In Richard Swinburne, editor, *La justificación del razonamiento inductivo.* Alianza, Madrid, 1976.

[286] Bertrand Russell. *Historia de la Filosofía Occidental.* Espasa Calpe, Madrid, 1997.

[287] Bertrand Russell. *Mysticism and logic*. Spokesman Books, 2007.

[288] Andrew M. Ryan. *The Substance of Spacetime: Infinity, Nothingness and the Nature of Matter*. Gadfly LLC, Virginia, 2 edition, 2016.

[289] W. C. Salmon. *Zeno's Paradoxes*. Hackett Publishing Company, Inc, Indianapolis, Cambridge, 2001.

[290] Wesley C. Salmon. Introduction. In Wesley C. Salmon, editor, *Zeno's Paradoxes*, pages 5 – 44. Hackett Publishing Company, Inc, Indianapolis, Cambridge, 2001.

[291] Steven Savitt. Being and Becoming in Modern Physics. In Edward N. Zalta, editor, *The Stanford Encyclopedia of Philosophy*. The Stanford Encyclopedia of Philosophy, 2008.

[292] E. Schrödinger. *What is Life*. Cambridge University Press, 1967.

[293] Erwin Schrödinger. *La naturaleza y los griegos*. Tusquets, Barcelona, 1996.

[294] Bruce A. Schumm. *Deep Down Things. The Breathtaking Beauty of Particle Physics*. The Johns Hopkins University Press, Baltimore, 2004.

[295] Jan Sebestik. La paradoxe de la réflexivitédes ensembles infinis: Leibniz, Goldbach, Bolzano. In Françoise Monnoyeur, editor, *Infini des mathématiciens, infini des philosophes*, pages 175–191. Belin, Paris, 1992.

[296] Manuel Sellés and Carlos Solís. *La Revolución Científica*. Síntesis, Madrid, 1994.

[297] Raymond A. Serway and Clement J. Moses aand Curt A. Moyer. *Modern Physics*. Thomson Learning, Inc., 2005.

[298] Claude Elwood Shannon. A Mathematical theory of communication. *The Bell System Technical Journal*, 22:379–423, 1948.

[299] Burra G. Sidharth. Comments on the mass of the photon, 2006.

[300] Z. K. Silagadze. Zeno meets modern science. *Philsci-archieve*, pages 1–40, June 2005.

[301] Hourya Sinaceur. ¿Existen los números infinitos? *Mundo Científico (La Recherche)*, Extra: El Universo de los números:24 – 31, 2001.

[302] Lee Smolin. *The Life of the Cosmos*. Phoenix, London, 1998.

[303] Lee Smolin. *Three roads to quantum gravity. A new understanding of space, time and the universe*. Phoenix, London, 2003.

[304] Lee Smolin. Átomos del espacio y del tiempo. *Investigación y Ciencia*, (330):58 – 67, Marzo 2004.

[305] E. Sober. Mathematics and indispensability. *Philosophical review*, 102(1):35–57, 1993.

[306] M. C. Solaeche. La Controversia entre L. Kroneckery G. Cantor acerca del Infinito. *Divulgaciones Matemáticas*, 3(1/2):115–120, 1995.

[307] Carlos Solís and Luis Sellés. *Historia de la ciencia*. Espasa Calpe, Madrid, 2005.

[308] Glenn Stark. Light. In *Encyclopedia Britannica*. Encyclopedia Britannica, 2021.

[309] Paul J. Steinhard. The inflation debate: Is the theory at the heart of modern cosmology deeply flawed? *Scientific American*, 304(4):18–25, 2011.

[310] Wolfgang Steinicke. Einstein and the Gravitational Waves. *Astronomische Nachrichten*, 326(7):640–641, 2005.

[311] Victor J. Stenger. *The comprehensible cosmos*. Prometheus Books, New York, 2006.

[312] Leónard Susskind. Los agujeros negros y la paradoja de la información. *Investigación y Ciencia (Scientifc American)*, (249):12 – 18, Junio 1997.

[313] José Manuel Sánchez-Ron. *El origen y desarrollo de la relatividad*. Alianza Universidad, Madrid, 1985.

[314] Eusebio Sánchez Álvaro. *LOS INGREDIENTES SECRETOS. Materia Oscura, Energía Oscura y las Nuevas Ideas sobre el Universo*. Cultiva Libros, Madrid, 2015.

[315] N. A. Tambakis. On the Question of Physical Geometry. In Franco Selleri, editor, *Open Questions in Relativistic Physics*, pages 141–147. Apeiron, Montreal (Canada), 1998.

[316] René Taton. *Historia General de las Ciencias*, volume 4 y 5. Orbis, Barcelona, 1988.

[317] Edwin F. Taylor and John Archibald Wheeler. *Spacetime physics. Introduction to special relativity*. W. H. Freeman and Company, New York, 1997.

[318] Richard Taylor. Mr. Black on Temporal Paradoxes. *Analysis*, 12:38 – 44, 1951 - 52.

[319] James F. Thomson. Tasks and SuperTasks. *Analysis*, 15:1–13, 1954.

[320] Paul A. Tipler and Ralph A. Llewellyn. *Modern Physics*. W.H. Freeman and Company, New York, 2000.

[321] Paul A. Tipler and Gene Mosca. *Physics for sicentists and engineers*. W.H. Freeman and Company, New York, 2008.

[322] Liang-Chen Tu, Jun Luo, and George T. Gillies. The mass of the photon. *Reports on Progress in Physics*, 68:77–130, 2005.

[323] D. Turner. *The Einstein Myth and the Ives Papers*, chapter Part 1. Absolute motion. Hope Publishing House, Pasadena (California), 1979.

[324] D. Turner and R. Hazelett. *The Einstein Myth and the Ives Papers*. Hope Publishing House, Pasadena (California), 1979.

[325] Gabriele Veneziano. El universo antes de la Gran Explosión. *Investigación y Ciencia (Scientifc American)*, (334):58 – 67, Julio 2004.

[326] J. P. Vigier. New non-zero photon mass interpretation of the Sagnac effect as direct experimental justification of the Langevin paradox. *Physics Letters A*, 234:75–85, 1997.

[327] John Carl Villanueva. How Many Atoms Are There in the Universe? Universe Today, 2009.

[328] Gregory Vlastos. Zeno's Race Course. *Journal of the History of Philosophy*, IV:95–108, 1966.

[329] Gregory Vlastos. Zeno of Elea. In Paul Edwards, editor, *The Encyclopaedia of Philosophy*. McMillan and Free Press, New York, 1967.

[330] Woldemar Voigt. Über das Doppler'sche Princip. *Göttinger Nachr.*, (8):41–51, 1887.

[331] G. H. Von Wright. *Time, Change and Contradiction*. Cambridge University Press, Cambridge, 1968.

[332] Jearl Walker. *Fundamentals of Physics*. John Wiley and Sons, Inc., 2008.

[333] David Foster Wallace. *Everything and more. Acompact history of infinity.* Orion Books Ltd., London, 2005.

[334] John Watling. The sum of an infinite series. *Analysis*, 13:39 – 46, 1952 - 53.

[335] Jeffrey R. Weeks. *The shape of space.* Marcel Derkker INC., 2002.

[336] Eric W. Weisstein. Continuum. From MathWorld. a Wolfram Web Resource, 2022.

[337] H. Weyl. *Philosophy of Mathematics and Natural Sciences.* Princeton University Press, Princeton, 1949.

[338] Frank Wilczek. *The lightness of being. Big questions, real answers.* Allen Lane. Penguin Books. (Kindle ebook edition), 2009.

[339] Frank Wilczek. *Las diez claves de la realidad.* Editorial Planeta. (Kindle ebook edition), Barcelona, 2022.

[340] Gideon Yaffe. Reconsidering Reid's geometry of visibles. *The Philosophical Quarterly*, 52(209):602–620, 2009.

[341] F. J. Ynduráin. *Electrones, neutrinos y quarks.* Crítica, Barcelona, 2001.

[342] Mark Zangari. Zeno, Zero and Indeterminate Forms: Instants in the Logic of Motion. *Australasian Journal of Philosophy*, 72:187–204, 1994.

[343] Eberhard Zeidler, W. (with Hackbush, and H.R.) Schwarz. *Oxford Users' Guide to Mathematics.* Oxford University Press, 2004.

[344] Paolo Zellini. *Breve storia dell'infinito.* Adelphi Edicioni, Milano, 1980.

Alphabetical index